T0138327

A Good That Transcends

A Good That Transcends

How US Culture Undermines
Environmental Reform

ERIC T. FREYFOGLE

The University of Chicago Press
Chicago and London

The University of Chicago Press, Chicago 60637
The University of Chicago Press, Ltd., London
© 2017 by The University of Chicago
All rights reserved. Published 2017.
Printed in the United States of America

26 25 24 23 22 21 20 19 18 17   1 2 3 4 5

ISBN-13: 978-0-226-32608-5 (cloth)
ISBN-13: 978-0-226-32611-5 (paper)
ISBN-13: 978-0-226-32625-2 (e-book)
DOI: 10.7208/chicago/9780226326252.001.0001

Library of Congress Cataloging-in-Publication Data

Names: Freyfogle, Eric T., author.
Title: A good that transcends : how US culture undermines environmental
   reform / Eric T. Freyfogle.
Description: Chicago ; London : the University of Chicago Press, 2017. |
   Includes bibliographical references and index.
Identifiers: LCCN 2016034774 | ISBN 9780226326085 (cloth : alk. paper) |
   ISBN 9780226326115 (pbk. : alk. paper) | ISBN 9780226326252 (e-book)
Subjects: LCSH: Environmental degradation—Social aspects—United States. |
   Environmentalism—United States.
Classification: LCC GE150 .F749 2017 | DDC 304.2/80973—dc23 LC record
   available at https://lccn.loc.gov/2016034774

♾ This paper meets the requirements of ANSI/NISO Z39.48-1992
(Permanence of Paper).

For the committed members and staff of
Prairie Rivers Network

CONTENTS

During the summer of 2015 Pope Francis released his extended encyclical, *Laudato Si'*, "Praise Be to You," his effort, as he explained it, "to enter into dialogue with all people about our common home" (¶ 3). It was a timely letter, and of vital significance, even as it drew together themes and messages that other observers of our earthly plight had presented for years. The earth, the pope proclaimed, is our common home. The health, fertility, and flourishing of that common home, including its human members, is an objective good that transcends our existence and knowledge, in ways we cannot fully grasp. We are abusing and degrading this home and hurting one another, other life forms, and future generations in the process. And we do so not because we are inherently bad people and not chiefly because of our damaging technologies. We do so for reasons embedded in our culture, in the ways we perceive the world, value it and its future, and understand our place in it; by the ways we comprehend ourselves chiefly as preference-satisfying individuals, content to compete within and organize our affairs around political-economic systems that generate deadly costs, ecological and social.

Modern culture, the pope contends, is deeply flawed and urgently needs to change. "Many things have to change course," the pope asserts, if we are to come to terms with our errors and head down the path of flourishing life, "but it is we human beings above all who need to change." "A great cultural, spiritual and educational challenge stands before us, and it will demand that we set out on the long path of renewal" (¶ 202).

This book engages the issues and challenges that weave through the pope's encyclical, issues that have, for many decades now, occupied the attention of other wide-ranging critics and need to become more visible. It engages our environmental plight, not by cataloging particular troubling practices, but by similarly standing back from the present to consider the whole of things.

Environmental problems arise because our behaviors toward nature and one another are in some sense misguided. That behavior in turn is significantly shaped and guided by modern culture and by the institutions, social systems, laws, and other arrangements that embody and strengthen this culture. The kind of forward movement that Pope Francis yearns to witness would require—it does require, surely—that we become more aware of our culture and question it soberly. Reform efforts need to address the true root causes, not merely the superficial wrongdoing, which is to say they need to bring about significant cultural change. Culture shapes and reflects how we see, feel, think, and talk, as well as how we act. Within a reformed culture, and to help foster it, we shall need to see, feel, think, and talk in better ways, more suited for our circumstances, our plight, and our best hopes.

These issues are taken up here through a series of inquiries that probe modern culture and its manifestations, starting from various points of beginning (including, in chapter 4, the pope's encyclical). As the inquiry unfolds, chapter by chapter, various themes and observations emerge, recur, and become deeper and richer. Ultimately they are brought together into an overall challenge to modern culture and into a call for significant change. That call proposes a quite considerable shift in the dominant culture trajectory of the past 250 years, particularly in the United States. It proposes, as a means to improve our dealings with nature, a far different intellectual and moral template than the one that has guided progressive reform throughout America's history.

The book's opening chapters pull wisdom from the sizable bodies of writing of three of the nation's most far-sighted writers on nature and culture, instructive individually and even more so when their particular perspectives are pieced together. The initial chapter looks at the mature conservation thought of Aldo Leopold, widely acknowledged at his death in 1948 as the nation's leading conservationist and whose star would rise higher still with his posthumous work, *A Sand County Almanac and Sketches Here and There*. Leopold is most remembered for the final chapter of that book, entitled "The Land Ethic," and for his now-familiar claim that we ought to calibrate our dealings with nature in terms of what is ethically and aesthetically right as well as what is economically expedient. Leopold's *Almanac*, though, was only one of his many published writings. Further, he left in his desk drawers at his death at age sixty-one some of his most vital conservation messages, draft writings laced with ideas he had reached in his final years. Beyond his writings Leopold was much sought as a speaker on conservation issues. Only a few of his talks were published, then or later.

The study here probes Leopold's mature thought by turning to these late talks, which are well documented in the Leopold archives. Leopold delivered over 100 conservation presentations to varied audiences over his final years. In them he distilled the reform messages that he believed were most critical, the ideas and ways of thinking that he thought Americans most needed to absorb and act upon if they were to mitigate their misuses of nature. By drawing these talks together—the dozens of manuscripts and note cards along with the few published versions—it is possible to reconstruct what might be termed Leopold's last talk. From the key elements of this mature conservation talk, aimed at a broad public and backed up by his contemporaneous writings, there is much to glean.

Chapter 2 turns to the conservation writer Wendell Berry, by wide consent a leading conservation voice of the latter twentieth century. As we shall see, Aldo Leopold grounded his conservation thought in the ecological realities of interconnection and interdependence. He stood at the forefront of ecology, a still-new science, and possessed considerable skill in reading landscapes, learning how they functioned and how humans had altered that functioning. His ultimate conservation message put the land's interconnections and functioning front and center, even as he also highlighted how misuses of nature stemmed less from ecological ignorance than from other aspects of the culture of his day. Wendell Berry, in contrast, has approached the challenges of land misuse by starting with the hilly farm landscape of his northern Kentucky home and the diverse rural culture centered there. While sharing Leopold's fascination with nature, Berry has engaged it mostly as a farmer and as a poet, his two principal hats. Much like Leopold, Berry has been variously dismayed and outraged by abusive land practices and pained by the resulting enduring wounds. Berry's response has been to turn a critical eye toward the culture at work in his homeland, identifying what is flawed in it while highlighting what has seemed healthy. Out of his reflection have come calls for renewed virtue and for improved personal character. Out of it too has come a direct challenge to ideas of liberty and individual rights when these entitlements are understood, as they are too often, in ways that deny necessary connections among neighbors, between people and land, and within the land's interlaced parts. To the overall cultural inquiry Berry adds social and virtue-based considerations that Leopold identified but only partly explored even as Berry fails to offer a coherent vision of how cultural change, and ensuing institutional reforms, might come about.

The book's third chapter turns to the thought of David Orr, long-time head of the environmental studies program at Oberlin College and, before that, guiding light of an experimental green learning center in the Ozarks. A frequent speaker at conservation gatherings everywhere, Orr is perhaps best known for his insistence that we use more wisely and virtuously the physical stuff we take from nature and that we bring improvements to our built landscapes. He is a designer at heart, attracted to technology that could greatly reduce our ecological footprints. One central message for Orr is that we return to nature the products of nature's fertility, maintaining and enhancing that fertility, and that we recycle and reuse, continuously, those parts of nature (minerals, for instance) and human-made substances that nature itself could not degrade and reuse: as Orr terms it, cradle-to-grave thinking in the case of nature's organics and cradle-to-cradle treatment in the case of all else. Orr, too, is very much a cultural critic. Like Leopold and Berry he knows that bad practices stem from the ways people see the world, value it, and understand their place within it. Taken alone, Orr's work misses some key elements. When added to the efforts of others, his work helps generate an intellectual and moral whole that is much more than its parts.

Finishing off the studies of leading conservation voices is chapter 4, which takes up at some length the encyclical by Pope Francis, identifying its main claims and putting them in the context of other writings on nature and culture. The recurring themes and claims of the encyclical bear striking resemblance with major claims made by Leopold, Berry, and Orr, and for sound reasons. The encyclical overall is solidly grounded in science, ethics, and careful environmental thought and its basic stances largely overlap with those of secular critics. What is new and vital is the way he refines and forges the material into a penetrating critique and a passionate call for reawakening. For centuries and still today much of the most thoughtful writing on ethics and the human plight has made use of religious language. It would be senseless to push aside the pope's plea because of his high church position or its faith-based frame.

Overall *Laudato Si'* is highly critical of modernity due to its degradation of the planet and just as much because of the vast economic and power inequalities that exist among the world's peoples, ills that he views (as do others) as closely related. Yet throughout his 246 numbered paragraphs Pope Francis manages to weave a strand of optimism. We are special creatures, he tells us, possessed of unique powers to learn and reason and with adequate free will to correct our ways. If we so choose and in the fullness of time we can rise above our limitations.

The three chapters following these begin from rather different places, looking not to the writings of leading conservationists but instead to enigmas that have long been central to the talk and work of environmental reform. Chapter 5 probes the institution of private property rights in nature and the ways we think and could think about it. Much abuse of nature takes the form of bad land-use practices, largely unfolding on lands that are privately owned. Scratch the surface of any land-use controversy, particularly any push for changes in land-use laws and regulations, and one quickly comes upon strident calls to respect and protect private rights. It is a call that draws ready assent from people who hear it, even as they may also fret the ill effects of private land misuse. While the environmental movement has long been active in pushing for new land-use laws (for instance, protections for wetlands and wildlife habitat and limits on sprawl), it has spent precious little energy questioning the institution of private ownership. It has failed to give rise to new visions of ownership that respect the sound historic reasons for private property while insisting that owners use their land and resource holdings in ecologically responsible ways. Chapter 5 takes on this task. It is labor that overlaps noticeably with the cultural inquiries and criticisms of Leopold, Berry, Orr, and Pope Francis. Private ownership is much more than a legal arrangement. It is also an embodiment of contemporary culture, particularly its economic liberalism. To see why private property has been such a roadblock to much-needed environmental reforms is to focus light also on some of the cultural bases of that resistance. A well-crafted vision of responsible land ownership could help forge a broader new vision of responsible individual liberty.

The sixth chapter turns to a topic that has long fascinated environmentalists of all types, the matter of wilderness. Looking back, perhaps the first major achievement of the environmental movement (conservation, as it was then still termed) was the enactment of the Wilderness Act of 1964. Love of wilderness remained strong in ensuing years, even as representatives of extractive industries chafed at the resulting limits on timber harvesting, grazing, and road-building. By the 1980s resistance to wilderness designations had become vocal and potent. For various reasons, the idea of wilderness and the rationales for leaving it untouched had by then become confused and contentious. Wilderness, it was being pointed out, was hardly a sensible land-use ideal for managing lands where people lived and worked. If wilderness was to serve as the benchmark for good land use, then any human alteration of nature was abusive, and the less of it, presumably, the better. Wilderness defenders, in short, were under attack on multiple fronts. Wilderness increasingly seemed like a luxury, beneficial mostly to wealthy recreationists who, as one critic put it, had no need to work for a living.

Both as physical place and cultural emblem wilderness is worth revisiting in the effort to locate the sources of today's impasses and to chart a pathway that offers hope. If we can think clearly about wilderness—what it is and why we might rightly protect it—then we are likely to think better about the broader roots of degradation and the cultural sources of today's resistance to change. Once clear on wilderness, we might do better casting a vision of a more life-enhancing culture. In any well-grounded vision wilderness preservation will play key roles, due not just to the benefits and values promoted by and within wilderness enclaves but also to their essential roles in promoting sound land uses on the larger landscapes that they help compose.

The seventh chapter takes as its starting point the much-discussed story of the tragedy of the commons, as recounted and examined by biologist Garrett Hardin in 1968. In his now-famous article, Hardin mostly spoke about the stresses of human overpopulation but it was his simple story of an overused grazing commons that caught widespread attention. His fictional tale was about a grazing landscape that was uncontrolled by anyone and that competing grazers could use as they saw fit, adding more animals at will. Such an open-access commons, Hardin contended, would inevitably slide downward from overuse unless the grazers took steps to limit their separate actions through one or another means of mutual coercion.

Hardin's tale has been grabbed onto over the years by an array of scholars and other commentators on people and land. From it they have drawn widely varied conclusions about the origins of land abuse and how it might best be addressed. Their varied approaches themselves invite a reconsideration of the simple grazing story, another look at it in light of writings by Leopold, Berry, Orr, and others on the cultural origins of ecological decline. Hardin's simple tale on its own is enigmatic, a trait that helps account for these widely conflicting commentaries. Only by digging deeper, only by probing in a direction that Hardin identified but did not pursue, can we get to the true causes of land degradation and give his tale, as it deserves, a new, more accurate title. Hardin had a sound sense of what was needed to avoid the tragedy but he did not recognize, as we must today, that the forces propelling degradation also feed resistance to the necessary reforms. New land- and resource-use limits are needed to avert tragedy. They in turn can arise only by reforms that address deeper, cultural flaws.

Of the issues and claims that weave throughout the chapters, four overlapping ones are usefully isolated up front:

First is the matter of how we perceive the natural order—too often in fragmented terms that dwell on specific parts of nature while ignoring or discounting interconnections, interdependencies, and emergent properties. Related to this is our constrained methodology for gaining understanding and our struggles to admit and accommodate sensibly the limits on what we know and can know.

Second, there is our tendency to comprehend ourselves chiefly as autonomous individuals—the upshot, we might say, of our long-rising liberal tradition. When we hold high the individual, exalting and respecting individual choice, we find ourselves less able to talk coherently about shared values or to distinguish them from personal preferences. We lessen our capacities and curtail our opportunities to deliberate collectively over the common good and then to act.

Third, there is the related question of value in the world, where it ought to reside and how we might properly acknowledge it: value in other life forms, in natural communities and processes, and in future generations. Were we to spread value more widely, diminishing our sense of human exceptionalism, how might we recast our hopes and senses of place? If we saw greater value in interconnections and social and natural communities as such, how might it change our visions and goals?

Then there is the need for us to gain or regain the language and ability, in concert, to talk seriously about how we ought to live in our natural home, that is, to engage soberly the normative aspects of our interconnected lives and to treat the topic as public rather than merely private business.

The final chapter reprises and draws together the central observations of the earlier ones. Its hope-filled aim is to shed light on a path that heads forward, toward more healthy lands and waters—remedies for our overstressed homelands—and toward ways of living that might just endure. It highlights how radical the needed cultural change would be and for nearly all people, not just for those who deny problems. The required cultural changes, it urges, in fact collide rather frontally not just with cultural elements insensitive to nature but also with the dominant template of progressive reform of recent generations. Committed as it is to the primacy of individual liberty and equality, today's liberal, progressive view embraces cultural elements that, in our dealings with nature, are in need of some pretty serious pruning.

The path ahead, as the final chapter proposes, needs to involve thinking and talking about our common home in much different ways. Efforts to promote environmental reform, accordingly, need to dwell above all on that cultural challenge. The public audience for it, for new ways of seeing, valuing, and talking, could be more receptive than we imagine.

ONE

Leopold's Last Talk

The career of conservationist Aldo Leopold took an important turn in the 1920s when he moved from the American Southwest with its expansive public lands to central Wisconsin, a region of fragmented land parcels mostly in private hands.[1] The arid Southwest was more ecologically sensitive than Wisconsin and its scars of human land abuse were more vivid. Yet Wisconsin too was a place where, to the trained eye, humans were failing at what Leopold termed "the oldest task in human history: to live on a piece of land without spoiling it."[2] The challenge in Wisconsin, as Leopold saw things, was to find mechanisms to compel, induce, or cajole private landowners to use their lands more conservatively, in ways that kept the lands fertile and productive for generations. For the next quarter century—until his death in 1948—Leopold searched for ways to meet that challenge, in the process digging more deeply into the human plight in nature than any American before him, and perhaps since.

In his many writings Leopold probed all aspects of that broad cultural and ecological movement then known as conservation, paying special attention to the sagging plight of private farms and farm landscapes. Particularly over his last decade he also delivered numerous conservation talks to varied audiences, a handful of them published (then or later), the vast majority not. So diligent was Leopold in retaining notes and manuscripts that we can reconstruct the main elements of some 100 of his talks from this period, when he spoke with greatest understanding and authority.[3] Leopold is best remembered for his literary gem, *A Sand County Almanac and Sketches Here and There*, a flowing, complex inquiry into the human role in nature, ecologically and philosophically.[4] In important ways, though, Leopold's mature conservation thought is most readily grasped by studying his oral presentations. It was in his talks that Leopold cut to the chase, reducing the complexity and ambiguity, curtailing his illustrations, and presenting his claims most directly.

By studying the records of these many talks it is possible to distill what might be termed Leopold's last conservation talk: not a specific talk given on a particular day to a particular audience but, even better, a talk constructed from shared elements of many talks, a generic conservation talk that expressed the points the mature Leopold deemed most vital. What were the messages that he emphasized, again and again, when he spoke to people about conservation? What were his key "take-home" points?

Over four decades of study and reflection Leopold came to understand how and why people misused land and what needed to change for them to behave better. His message was at once radical and conservative. And even as it built upon the best science, the message he offered chiefly had to do with human perceptions, cultural values, and the social institutions and practices built upon them. Leopold is much cited today, yet his message as often popularized is greatly muted, to a claim that he mostly proposed trial-and-error land management or that we simply "be nice to nature." His true message had a much sharper bite, and it went well beyond criticizing specific land-use ills.

Leopold was an intellect of considerable depth as well as breadth. Slowly, carefully, he rested his conservation basics and scientific understandings on a well-considered reassessment of how humans fit into nature and how they might best understand and embrace their ecological plight. In the end, after decades of practice, study, and reflection, Leopold called Americans to make profound changes to the nation's longstanding traditions of individual autonomy and economic liberty and, indeed, even to main components and dualities of Enlightenment thought. Only change at such fundamental levels, Leopold reluctantly concluded, could allow human life to flourish. Only by becoming different and better in our understandings, ethics, and aesthetics, only by accepting a more humble status and undergoing (as he put it in 1941) a "face-about in land philosophy," could we flourish while sustaining other life forms and processes.[5] "Thus we started to move a straw," he explained to fellow wildlife professionals in a 1940 talk, "and end up with the job of moving a mountain."[6]

Leopold was critical of the conservation of his day, particularly conservation education that was, he contended, a "milk and water" affair, far too timid and unimaginative to prompt fundamental change.[7] Alive today, he might well say the same about the fragmented, technical, narrowly focused work of the contemporary environmental movement. It similarly fails to identify the root causes of land abuse in human nature and culture. Failing to see them, failing to dig that deeply, the movement lacks a strategy that inspires hope.

The Talk

The conservation community of Leopold's era, from about 1900 to the period after World War II, aspired above all to redress the specific resource challenges identified at the turn of the prior century, the problems of declining flows of those parts of nature—those natural resources—that humans used directly. Since the late colonial era croplands had declined in natural fertility and without inputs produced lower yields. Game populations were sliding down while fishers and whalers journeyed ever further to find their prey. Timber clearcutting appeared to threaten flows of wood products; industrial processes and human wastes tainted water supplies. Agriculture, it seemed, could expand only by draining rivers and drawing down aquifers. Dust storms in semi-arid lands—and even normal rainfall on hillsides— almost inevitably reduced valuable topsoil into unwanted sediment, clogging rivers and reservoirs. The typical fear-driven solutions of the era sought to manage resource flows more scientifically. Yet problems remained, particularly as steps to conserve one resource clashed with measures taken to protect and produce others. Meanwhile, attentive observers recognized that active efforts to enhance annual flows of specific natural resources came at great cost both to the countless wild species that were simply in the way and to the ecological processes and natural beauties that they sustained. Underlying and justifying this scientific, resource-conservation effort were key assumptions about human powers and science, about the moral primacy of human life, and about the economic and political importance of individual autonomy.

This was the intellectual and moral environment in which Leopold came of age, rose through the institutional (Forest Service) and professional ranks, and gained prominence as a forester, game manager, wilderness advocate, and penetrating writer. It was also the cultural milieu that Leopold confronted when late in life he reached out to varied audiences to talk about the nation's conservation needs. However consciously, his audience members assumed that moral value resided largely if not entirely in the human species and that humans were best understood as more or less independent, autonomous beings. Similarly, nature for them existed largely as a warehouse of raw materials and seemed created precisely for that purpose. Guided by human cleverness, science and industry supplied the tools for extraction and manipulation, solving problems as they arose. Landscapes were divided among political jurisdictions and, in most of the country, into clearly bounded land parcels, privately owned and managed. The rights of private landowners were substantial and somehow grounded in the constitution and individual rights. Limits on their land-use options were legitimate only when private actions caused visible, immediate harm to neighbors or the surrounding community.

By his mature years, Leopold came to believe that this entire constellation of perceptions and values lay at the root of America's environmental plight. Bad land use was intertwined with these cultural components, and it would end only if and when American culture changed directions. Thus, as Leopold rose to address his audiences his central, ambitious aim was to push American culture in a new, healthier direction. He did so by emphasizing, above all, four central messages about the land as a community of life, the proper or healthy functioning of that community, the prudence and virtue of embracing community (or land) health as an overall conservation goal, and the extraordinary challenge humans faced in pursuing that goal.

The land as community. Leopold's first hope in his standard conservation talk, logically if not always temporally, was to push his audience to think in new ways about land and the human place in the land. Land was not simply a warehouse or flow of resources that humans needed in order to live. To the contrary, the land—understood as not just soils and rocks but water, plants, animals, and people—was a highly integrated, interdependent functioning system upon which all life depended for survival, human life included. "Before I even begin," Leopold explained to one audience, "I must ask you to think of land and everything on it (soil, water, forests, birds, mammals, wildflowers, crops, livestock, farmers) not as separate things, but as parts— organs—of a body. That body I call the land (or if we want a fancy term, the biota)."[8] This land was the "most complex" of all organisms, he told a campus group in May of 1941.[9] "No one dreamed a hundred years ago that metal, air, petroleum, and electricity could coordinate as an engine," Leopold explained in 1939. "Few realize today that soil, water, plants, and animals are an engine, subject, like any other, to derangement," a "biological engine" that had to be used not just with skill but with enthusiasm and affection.[10] As he wrote on a 3×5 lecture notecard prepared around 1942: "Land: soils, water, plants, animals."[11]

Leopold frequently used various metaphors to explain this view of nature. A common one, particularly when talking about ethics and perceptions, was to speak of land as a community, a term that skirted some of the imprecisions of describing it as either an organism or a mechanism.

The land was a community, and humans were as integrated with its other components as any other living creature. As Leopold would famously put it in his *Sand County Almanac*, "We abuse land because we regard it as a commodity belonging to use. When we see land as a community to which we belong, we may begin to use it with love and respect." His land ethic, he explained, in effect changed "the role of Homo sapiens from conqueror of the land-community to plain member and citizen of it."[12] "Who is the land?" he asked rhetorically in 1942. "We are, but no less the meanest flower that blows. Land ecology discards at the outset the fallacious notion that the wild community is one thing, the human community another."[13]

Regrettably, Leopold lamented, this understanding of land was simply not widely understood. "We have taught science for a century," he complained, "without implanting in the mind of youth the concept of community with the land."[14] Conservation simply could not succeed until people saw the land in this new way. There was "only one way out of this confusion": "For the average citizen to have a wider appreciation of land, a more critical understanding of it, especially his own land."[15] The underlying educational challenge, he understood, was a huge one. "We find that we cannot produce much to shoot," Leopold said to the Wildlife Society, "until the landowner changes his way of using land, and he in turn cannot change his ways until his teachers, bankers, customers, editors, governors, and trespassers change their ideas about what land is for. To change ideas about what land is for is to change ideas about what anything is for."[16]

A community can be more or less healthy. Leopold spent years of study and reflection attempting to learn how the land community functioned and how people might best evaluate the quality or condition of their lands. The key perceptual step was to see that land was not simply a collection of constituent parts, however complex. To the contrary, land's components were sufficiently interdependent that failings in one part of the land community could undercut the productivity of other parts. Leopold addressed this issue in a talk to wildlife professionals in 1939 as he surveyed the intellectual gains of the past decade. "The greatest single gain since 1930 lies, I think, in the growth of detail in the idea that resources are interdependent. We knew then that you can't have healthy fish in sick waters. We knew something of the interdependence of animals and forests. But the idea of sick soils undermining the health of the whole organic structure had not been born."[17] He continued the theme in 1947 when summing up a major wildlife conference: "This Conference has reaffirmed, through half a dozen speakers, the basic interdependence of soil, water, crops, forests, wildlife, and people. We conservationists have heard this before, but let no man forget that the average American has not heard it yet." "Conservation is usually thought of as dealing with the *supply* of resources," Leopold stated in 1944. "This 'famine concept' is inadequate, for a deficit in the supply in any given resource does not necessarily denote lack of health, while a failure of function [arising from disorganization of the land] always does, no matter how ample the supply."[18]

It was a substantial, long-term struggle for scientist Leopold to gain a sense of what it meant for a land community to possess health. Starting in 1935 he began listing what he termed the main signs of land sickness or pathology. "Regarding society and land collectively as an organism," he announced in 1935, "that organism has suddenly developed pathological symptoms, i.e., self-accelerating rather than self-compensating departures from normal functioning."[19] It was only years later that Leopold was willing to convert his evidence of land sickness into a positive, albeit generalized, definition of land health. One expression of it came in a 1944 manuscript first published in 1991:

> The land consists of soil, water, plants, and animals, but health is more than a sufficiency of these components. It is a state of vigorous self-renewal in each of them, and in all collectively. Such a collective functioning of interdependent parts of the maintenance of the whole is characteristic of an organism. In this sense land is an organism, and conservation deals with its functional integrity, or health.[20]

One of Leopold's fullest expressions of land health appeared in a draft document prepared not long before he died, perhaps intended as the text for a major address he was slated to give as outgoing president of the Ecological Society of America, some months after his premature death.[21]

> The symptoms of disorganization, or land sickness, are well known. They include abnormal erosion, abnormal intensity of floods, decline of yields in crops and forests, decline of carrying capacity in pastures and ranges, outbreak of some species as pests and the disappearance of others without visible cause, a general tendency toward the shortening of species lists and of food chains, and a world-wide dominance of plant and animal weeds.[22]

In talk after talk, Leopold stressed that the land's functioning as a community could be more or less sound, more or less healthy, and that its productivity and thus capacity to sustain life was based on that health. Leopold lacked full confidence in his own understanding of land health—far more work was needed—and encouraged others to join in his quest to make sense of it. Indeed, he was sometimes prone to present the norm as a question: "What is land-health?"[23] Yet he knew well enough the major symptoms of sickness, and he possessed plentiful evidence that sick lands were less able to sustain human communities.

Land health as the conservation goal. The first two points that Leopold presented in his standard talk—that land was a community and that the community could be more or less healthy—led directly to his third point: the health of the land should be the aim of all conservation efforts. This normative claim, Leopold knew, ran counter to the accepted wisdom of the age, which focused on sustained flows of discrete resources. "The basic fallacy in this kind of 'conservation' is that it seeks to conserve one resource by destroying another," Leopold told a garden club in 1947. "These 'conservationists' are unable to see the land as a whole. They are unable to think in terms of community rather than group welfare, and in terms of the long as well as the short view."[24]

Again and again Leopold repeated his complaint against the conservation ideology of his day. "We have hundreds of conservation organizations, each promoting some special resource, often at the expense of another," he lamented in 1939. "None sees land as a whole."[25]

"Conservation is more than commodities," he reiterated the next year; "the various kinds of commodities shouldn't compete, [they] should be complementary."[26] By focusing on specific conservation challenges, "we confuse the symptom and the disease, the part and the whole."[27] Given the frequent clashes among them, the conservation technologies of the day were simply not working even though their practitioners tried to coordinate their efforts. "They lack, firstly, a collective purpose: stabilization of land as a whole," Leopold explained. "Until the technologies accept as their common purpose the health of the land as a whole, 'coordination' is mere-window dressing, and each will continue in part to cancel the other."[28]

"Basic to all conservation is the concept of land-health; the sustained self-renewal of the community," Leopold explained to a wildlife group in 1941. "It is at once self-evident from such an over-all view of the community that land-health is more important than surpluses or shortages in any particular land-product."[29] It was thus essential that "sound conservation propaganda . . . present land health, as well as land products, as the objective of 'good' land use."[30] Or as he put it in the outline for one talk, "Conservation—health of land."[31]

Leopold made clear his emphasis on land health in the fall of 1946 when he was asked to draft the conservation platform for a fledgling national political party being organized by John Dewey and A. Philip Randolph.[32] Leopold responded with a conservation platform that fit easily on one page so that its main points might stand out:

> The health of the land as a whole, rather than the supply of its constituent "resources," is what needs conserving. Land, like other things, has the capacity for self-renewal (i.e., for permanent productivity) only when its natural parts are present, and functional. It is dangerous fallacy to assume that we are free to discard or change any part of the land we do not find "useful" (such as flood plains, marshes, and wild floras and faunas).[33]

Radical change. By this point in his standard talk, Leopold had made three of his four key points. The land was a community in which humans were embedded; that community could be more or less functionally efficient and fertile, which is to say healthy; and the health of the community as such, not the flows of particular "resources," should be the overriding aim of conservation. What remained was to make his most difficult and sensitive point, to explain to people—without alienating or scaring them—the kind of radical change required in American culture for humans to live on land without spoiling it. His message on this point, in truth, called for a redirection of the trajectory of Western culture since the era of Descartes and Francis Bacon in the early Enlightenment—a turning away from popular forms of liberal individualism and a reassessment of the achievements and possibilities of science and the scientific method. This was not a message that Leopold could present directly, in the language of philosophy or political theory. Instead he had to simplify his message in some way, translating it for audiences into ordinary language and into everyday life. Leopold did so by emphasizing the need for people to embrace not just new ideas as such, but new feelings, new values, and new goals.

As he warmed to this issue Leopold often criticized pointedly the popular mind-frame of his day. "Land, to the average citizen," he complained, was "still something to be tamed, rather than something to be understood, loved, and lived with. Resources are still regarded as separate entities, indeed, as commodities, rather than as our co-inhabitants in the land-community."[34] As he put it in a wartime presentation:

> Land, to the average citizen, means the people on the land. There is no affection for or loyalty to the land as such, or to its non-human cohabitants. The concept of land as a community, of which we are only members, is limited to a few ecologists. Ninety nine percent of the world's brains and votes have never

heard of it. The mass mind is devoid of any notion that the integrity of the land community may depend on its wholeness, that this wholeness is needlessly destroyed by the present modes of land-use, or that the land-sciences have not yet examined the possibilities of preserving more of it.[35]

A key flaw in the popular mind was the assumption that humans somehow stood apart from nature and could manipulate it at will, overcoming challenges as they arose:

Conservation is a pipe-dream as long as *Homo sapiens* is cast in the role of conqueror, and his land in the role of slave and servant. Conservation becomes possible only when man assumes the role of citizen in a community of which soils and waters, plants and animals are fellow members, each dependent on the others, and each entitled to his place in the sun.[36]

At the center of the popular misunderstanding was America's love affair with an industrial system that treated nature simply as a fund of raw materials. "It is increasingly clear," Leopold asserted, "that there is a basic antagonism between the philosophy of the industrial age and the philosophy of the conservationist."[37] Or as he put it in a letter to fellow wildlife researcher Bill Vogt, commenting on Vogt's conservation ideas: "The only thing you have left out is whether the philosophy of industrial culture is not, in its ultimate development, irreconcilable with ecological conservation. I think it is."[38]

What was needed was a new orientation of people to land, one that grew in the heart as well as the mind. "Culture is a state of awareness of the land's collective functioning," Leopold observed in 1942, and a better culture was urgently needed, one based on "a wider appreciation of land, a more critical understanding of it."[39] In other words, "the basic question in conservation [was] not the condition of the land, but the proportion of people who love it."[40]

There must be some force behind conservation more universal than profit, less awkward than government, less ephemeral than sport; something that reaches into all times and places, where men live on land, something that brackets everything from rivers to raindrops, from whales to hummingbirds, from land estates to window boxes.

I can see only one such force: a respect for land as an organism; a voluntary decency in land-use exercised by every citizen and every landowner out of a sense of love for and obligation to that great biota we call America.[41]

In many of his presentations Leopold paid particular attention to the category of citizens who were most vital if America was going to see land anew. Most land in Wisconsin was owned and controlled by farmers, and it was farm culture above all that required significant change. It was essential that farmers develop a new understanding of what it meant to use farmland well and to succeed as a farmer:

> In addition to healthy soil, crops, and livestock, [the farmer] should know and feel a pride in a healthy sample of marsh, woodlot, pond, stream, bog, or roadside prairie. In addition to being a conscious citizen of his political, social, and economic community, he should be a conscious citizen of his watershed, his migratory bird flyway, his biotic zone. Wild crops as well as tame crops should be a part of his scheme of farm management. He should hate no native animal or plant, but only excess or extinction in any one of them.[42]

This new attitude toward land, Leopold believed, had to come together in moral terms, as a matter of right and wrong, not merely in the untethered language of individual preference or desirability. As he put it to a garden club, they should not as advocates shy away from moral admonition:

> The direction is clear, and the first step is to *throw your weight around* on matters of right and wrong in land-use. Cease being intimidated by the argument that a right action is impossible because it does not yield maximum profits, or that a wrong action is to be condoned because it pays. That philosophy is dead in human relations, and its funeral in land-relations is overdue.[43]

The conservation message most popular at the time was simply too easy to get much done, he asserted. "It calls for no effort or sacrifice; no change in our philosophy of values"; it failed to recognize that "no important change in human conduct [was] ever accomplished without an internal change in our intellectual emphases, our loyalties, our affections, and our convictions."[44]

Situating Leopold's Claims

So lyrical is Leopold's writing, particularly in the *Almanac* and other polished works, that his words and phrases sweep readers along without insisting that one go slow and reflect. Few readers, then or even now, paused to consider how radically Leopold sought to reshape modern culture. Few could see that Leopold aimed not to prune unhelpful cultural shoots but rather to pull society up by its roots and replant it in better soil, more moral and intellectual.

By late in life Leopold had grave reservations about Western civilization and the idea of progress. The Western trajectory featured a mixed heritage of darkness, decay, and violence as well as enlightenment and elevation. In too many ways humans were blind and arrogant. Like civilizations of the past, the modern world was degrading its natural foundations and thus its future. Its cleverness in developing tools and harnessing power far surpassed its advances in ethics and aesthetics.

Leopold's messages gain complexity when we situate his thinking within influential strands of philosophy over the centuries, not so much to trace actual influences on him, but to highlight, clarify, and evaluate his central challenges. One can do so by assessing where Leopold situated himself on five subjects of enduring interest to philosophy:

- How distinct are humans from other life forms and are they sensibly understood, as the liberal tradition would have it, as autonomous, rights-bearing individuals; is human arrogance, that is, consistent with scientific reality?
- Is science, as assumed, on the verge of understanding nature and controlling it; is our cleverness, that is, sufficient to overcome the limits on our senses and knowledge?
- Is nature largely a collection of parts—some valuable to humans, most not—and can we rightly think of nature and deal with it in terms of its parts?
- Is humankind well guided by embracing a scientific understanding of truth, given the vast gaps in human knowledge?
- Is there, in the physical world, an overriding norm of goodness that humans ought to respect or can we, given the collapse of faith-based verities, sensibly equate goodness with the satisfaction of human preferences?

Leopold tangled with these big issues, drawing conclusions that set him far from the dominant views of his day.

Human exceptionalism and liberal autonomy. Perhaps Leopold's central challenge to modernity had to do with his ultimate understanding of the human place in nature, a place that was, he concluded, far more humble than we recognized. On this issue Leopold relied in important part on modern science, which increasingly cast doubt on the presumptions of liberal humanism.

The prevailing understanding of Leopold's day rested on a centuries-long intellectual trajectory, one that gained prominence at the Enlightenment's dawn in early seventeenth-century Western society. The then-ascending impulse was for humans to rise above nature, seeing nature as a complex but ultimately knowable machine and controlling it in service of human wants. It was an impulse—grounded on the humanist side in the Renaissance— that gave rise in complex ways not only to advances in science and technology but to the revolutions of the seventeenth and eighteenth centuries, the emergence of economic liberalism, and the expanding embrace of individual rights. Put simply, the independent thinker of the age of Descartes (early seventeenth century) had matured into the morally autonomous, utility-seeking actor of the age of Bentham and J. S. Mill (nineteenth century) and gone onward to become the rights-bearing, vote-wielding citizen of the early twentieth century. In the emergent liberal ideal, an individual could act as she saw fit, crafting and pursuing a self-chosen vision of the good life, so long as she caused no material harm and recognized the equal rights of others to act similarly. Nature was where human life unfolded, and science helped guide its manipulation. Driving the quest, as historian Richard Tarnas has observed, was "a heroic impulse to forge an autonomous rational human self by separating it from the primordial unity with nature."[45]

The problem with this trajectory, as Leopold well understood, was that the very science that made humans proud and powerful had begun, by the mid-nineteenth century, to cast doubt on these assumptions of human exceptionalism. Theories of evolution and natural selection questioned the uniqueness of human life; we differed from other life, it seemed, not in kind but merely in degree. Freudian psychology questioned whether humanity was in fact guided by reason rather than, like other creatures, animal passions. Meanwhile, and more important, claims of objective morality and goodness—particularly the religious ones that exalted humans as a special life form—were rapidly losing their potency. Was the reigning moral order with humans on top simply a human conceit? And was the Western world's particular worldview, as anthropologists and sociologists pointed out, merely one of countless worldviews that humans had embraced at different times and places—mere human creations all of them, lacking in objective reality?

As Leopold considered the human place in nature he drew heavily upon the latest science. Evolution supplied the base of his worldview; humans arose in the same way as other species and thus shared ancestry and genes. As much or more, though, he was influenced by the newer field of ecology, which by focusing on present-day interdependencies operated, in a sense, perpendicular to the temporal flow of evolution. Much as Darwin forged a historic link between humankind and other life forms, so ecology showed that humans today were every bit as connected and interdependent with nature as the lowly earthworm. And it was a connection that was present-day and essential, not merely, like evolution, a story of eons past. Ecology portrayed an ever-changing natural order upon which all life depended. As Leopold put it to a student audience in 1941:

> Every living thing represents an equation of give and take. Man or mouse, oak or orchid, we take a livelihood from our land and our fellows, and give in return an endless succession of acts and thoughts, each of which changes us, our fellows, our land, and its capacity to yield us a further living.[46]

Ecology, Leopold understood, was no less powerful than evolution in questioning the arrogance of the Western liberal view, in challenging its presumption of human specialness and its tendency to portray humans as free-standing individuals. To the contrary, as science showed, the human being was in important ways simply another animal that lived, ate, reproduced, and died. Humans, too, were merely components of something larger and could not be understood without considering their interactions with natural systems and other life forms.

Leopold's intellectual journey led him, step by step, to a radical reconception of the human place in nature. Conventional morality notwithstanding, the individual human was in physical fact embedded in a natural order that could be more or less conducive to life. Writing at about the same time, philosopher John Dewey stressed that individuals were embedded in society with much of what they understood and did guided by society. With this perspective Dewey carried forward the transcendentalist dislike of atomistic thinking; as his Vermont predecessor James Marsh asserted (echoing various ancient Greeks), people only realized themselves in and through communities and as they successfully filled social roles.[47] Leopold adapted this organic thinking to the natural world and modern science, serving in effect as Dewey's ecological counterpart. In Leopold's view, individuals were called to play ecological as well as social roles, particularly when they wielded the power to manage land, and they truly flourished only when they fulfilled their roles well.

In "The Land Ethic," the ultimate essay in his *Sand County Almanac*, Leopold expressed plainly his dissent from the Western liberal tradition. Far from being conqueror of the land community, the individual was simply a "plain member and citizen of it." He was, "in fact, only a member of a biotic team," and as such owed duties of responsible conduct to both the team and its other members.[48] In the classic liberal view of John Stuart Mill, the individual was free to act so long as he caused no harm to others.[49] But what did the do-no-harm limit mean when an individual belonged to a land community and when every action spread ripple effects far and wide? One could no longer define harm solely as direct impacts on human neighbors. Harm also occurred by the degradation of the community's ecological functioning, by disruptions to the health of the land as such. And it was no longer acceptable, Leopold implied, for a landowner to sit back and do nothing when land health required positive action.

The reach of human knowledge. Leopold's reconsideration of the human plight and human capabilities drew him into longstanding discussions about how much humans knew and could know, which is to say into the field of epistemology. Here, too, he developed a sense of humility that set him apart, even with his extraordinary grasp of modern science.

As Leopold studied the natural world, he had no doubt of its real existence, nor did he question that scientists could learn truth by using their senses to gather data and applying reason to the results. At the same time, though, Leopold accepted the view (often traced to Kant) that our knowledge of nature is constrained by limits on our senses and filtered through our brains, that our knowledge is necessarily interpretive, however much we strive to connect directly to things-in-themselves.[50] As the American pragmatists had put it, our knowledge of nature was not a matter of certainty but of greater or lesser degrees of confidence. And, as Leopold knew, confidence levels varied enormously.

Repeatedly in his talks Leopold sought to disabuse audiences of their assumptions about both science's accomplishments and its prospects. "The ordinary citizen today," he observed, "assumes that science knows what makes the community clock tick; the scientist is equally sure he does not. He knows that the biotic mechanism is so complex that its workings may never be fully understood."[51] "As a matter of fact," he commented on another occasion, "the land mechanism is too complex to be understood, and probably always will be. We are forced to make the best guess we can from circumstantial evidence."[52] And again: "The land-mechanism, like any other mechanism, gets out of order. . . . Science understands these disorders superficially, but it seldom understands why they occur. Science, in short, has subjugated land, but it does not yet understand why some lands get out of order, others not."[53] Leopold was particularly irritated by those who claimed to know which species were valuable and which could be lost without cost to people, a question that mixed scientific fact with normative judgment. Early ecologists may have embraced that conceit, Leopold acknowledged, but they were wrong:

Economic biology assumed that the biotic function and economic utility of a species was partly known and the rest could shortly be found out. That assumption no longer holds good; the process of finding out added new questions faster than new answers. The function of species is largely inscrutable, and may remain so.[54]

Yet even as he recognized these limits on human powers, Leopold realized that people had to act; they had to make decisions based on the conclusions they could draw. The sensible approach, he believed, was to employ our best science, even as we recognized its deficiencies and pushed for further research. Leopold was particularly insistent that scientists step forward, playing the role of citizen as well as expert, and offer their best professional judgment on what it took to sustain the land's health. Unless they did so, nonscientists who knew even less would take the lead. He made the point when ending a presentation on land health as an overall conservation goal:

> These then are my personal guesses as to the conditions requisite for land-health. Some of them step beyond "science" in the narrow sense, because everything really important steps beyond it. . . . Objectivity is possible only in matters too small to be important, or in matters too large to do anything about.[55]

Perhaps most striking in Leopold's humble comments on the capacities of science—humble given his enormous personal knowledge of plants, animals, and natural processes—was his frequent claim that scientific inquiry needed to be informed and inspired by sources outside it, particularly by arts and the imagination. Leopold illustrated his approach when describing an autumn landscape in the northwoods, a landscape that was not complete, in his view, without the presence of the ruffed grouse. The significance of the grouse, he asserted, was "inexpressible in terms of contemporary science." It arose because the grouse embodied an "imponderable essence" that philosophers termed the *noumenon*, an essence "in contradistinction to *phenomenon*, which is ponderable and predictable, even to the tossings and turnings of the remotest star."[56] As Leopold stressed to professional colleagues, he hoped that "this senseless barrier between science and art may one day blow away."[57]

An organic whole, or collection of parts? Leopold, as noted, explained repeatedly that nature exists as an organized community of interdependent, co-evolving life forms. In doing so, he strongly countered those who spoke of nature as a collection of parts, as a warehouse of resources for humans to manage as they saw fit. Leopold's organic view was hardly new; indeed, it represented perhaps the dominant perspective in all of human history. It bore similarities, for instance, with the German philosophic tradition—which had resisted the atomism of French and Anglo-American liberals—and with the views of ancient Stoics who, as one historian explained, understood the whole of the universe as ordered and animate:

> For the Stoics, the structure of the world—the cosmic order—is not merely magnificent, it is also comparable to a living being. The material world, the entire universe, fundamentally resembles a gigantic animal, of which each element—each organ—is conceived and adapted to the harmonious functioning of the whole.[58]

The Stoic tendency was to see this natural order as a perfect one. Leopold made no such claim of perfection, nor did he contend with Plato that nature's order was shaped and guided by "a wondrous regulating intelligence."[59] Closer to Leopold, then, were perhaps Romantics of the eighteenth and nineteenth centuries who, in more secular ways, stressed the organic wholeness of nature and its ineffable mystery.[60] Leopold's view, though, drew more on science and incorporated nature's dynamism, an awareness that species came and went and that biotic communities were inexorably pushed and rearranged by geological and climatic forces. The natural world that Leopold sensed was ever shifting, yet it was an interdependent functioning whole nonetheless.

Leopold in the early 1920s took an interest in the unusual philosophic writings of Piotr Ouspensky, a Russian philosopher-mystic whose major work, *Tertium Organum* (English translation 1920), contended that nature in its wholeness was infused with spirit and intelligence.[61] The mystic's assertions apparently resonated with Leopold, yet his own views may have been closer to those of Frenchman Henri Bergson, whose influential *Creative Evolution* dated from 1907. Bergson argued that evolution and thus the life-creating process was powered if not guided by a vital impetus (*élan vital*), a mysterious life force that pushed nature to ever higher forms of complexity.[62] Leopold did not embrace Bergson's thought overtly, or that of any other vitalist. Yet like Bergson he seemed at times unwilling to view nature in strictly material terms. Some force of some sort—perhaps Bergson's *élan vital*, perhaps something else—brought the physical stuff of nature to life, creating an organism that was more and other than the sum of its parts.

As he sought to make sense of nature Leopold was certainly not an explanatory reductionist; like most contemporaries he knew nature could not be explained by describing its parts in isolation. Much like the tissues that composed an organism, the parts of a land community gave rise in their interactions to emergent properties and forms of functioning that were not present in any of the parts in isolation. Yet at moments Leopold seemed to push beyond the emergentist stance, to suggest that nature had more than just physical parts and novel properties created by the interaction of those parts. There was something intangible if not spiritual also present, permeating and animating the whole.

Particularly in his late years Leopold opened himself to nature as fully as he could and invited others to do so also, listening to its music, absorbing its forms and colors, and imagining all that remained hidden from view. In this regard we might compare his work with Mark Twain's *Huckleberry Finn* and Twain's explorations of ways of knowing the river. Twain the former boatman knew the pilot's way of seeing the river, objectively attending to its physical moves and respecting its raw power. Yet there was also the way of passenger and poet, swept along with the river's majesty and beauty, for whom the river was more and other than a flow of water, more even than a community of life. Perhaps, in the end, this is what Leopold meant to suggest—not that nature in its wholeness contained intangible elements, but simply that it inspired awe, that it awakened in humans senses that could come from no other source.

Standards of truth, and the need to act. Leopold's awareness of human ignorance and the inevitable, endless limits on science led him predictably to a state of unease when it came to deciding whether to embrace a new fact about nature. The scientist in him doubtless wanted a high degree of proof, enough evidence to support a conclusion to a high level of confidence. Still, he seemed troubled by this perspective when it came to accepting evidence of our misuse of nature and particularly when crafting a normative vision of land health. It was evident enough that humans were sapping the health of landscapes. Remedial action was thus needed, urgently. Could that action wait until scientists had great confidence in their findings of land sickness? Could managers postpone deploying improved methods of land use until researchers had higher confidence in their benefits? In the research laboratory a high barrier to truth often brought benefits. But might lesser standards of truth be used in the face of widespread decline and the need to make immediate changes?

The scientific ideal of truth, the one Leopold would have absorbed from his studies, defined truth in terms of correspondence with physical reality: a fact about nature was true insofar as it mirrored the physical world, without human distortion. Leopold knew, however, that scientists worked incrementally, building upon facts that they accepted as true. The likelihood that a new, proposed fact was true therefore turned in part on whether it fit together with other facts that seemed true. This approach borrowed from a different definition of truth, one that tested a proposed fact in part based on whether it fit together sensibly with all else that was accepted as true. This second definition—the coherence theory of truth—was typically not an ultimate definition; it did not displace truth as complete correspondence. It was instead a more practical, expedient approach to truth in which it made sense to accept and act upon a fact that cohered with other truths, even as the quest continued for correspondence-based truth. Researchers with no other duties might get by with insistence on truth as correspondence. But land managers and conservationists were differently placed. Indeed, it could be costly to make decisions based solely on facts proven with very high reliability and overlooking the many gaps. Better land-use results would often come by accepting a less lofty level of proof.

Leopold understood that scientific understandings evolved over time by means of a process that was never-ending, as one set of ideas was augmented or displaced by another. "All history shows this," he told an engineering group in 1938, "that civilization is not the progressive elaboration of a single idea, but the successive dominance of a series of ideas." Like pragmatist Charles Sanders Peirce, Leopold gave weight in his definition of truth to the shared embrace of a particular conclusion; social acceptance was significant.[63] What Leopold had in mind, though, was not the popular sentiment of his de-natured fellow citizens generally. Like Peirce he had in mind the consensus views of his scientific colleagues: Science was a group effort that proceeded by fits and starts.

Two aspects of Leopold's attitude toward truth stand out most clearly, and they link him on this point to leading American pragmatists of the late nineteenth and early twentieth centuries. Leopold's research, as noted, was guided by a felt need to find ways for people to live on land without degrading it. It was purpose-driven, and the hour was late. The obvious approach was to act using the best current understandings, even as searches went on, not just for new knowledge, but to refine and replace conclusions that were tentatively accepted as true. To admit that ideas accepted as true today would be altered in the future did not undercut their comparative value today so long as they moved people in a useful direction, toward the normative goal of land health.

Pragmatists such as Peirce, William James, and John Dewey did not abandon the popular view of truth as accurate correspondence with reality; it remained the ideal that science pursued. But with perfection impossible it made sense in their view to assess competing ideas and assertions in terms of the consequences of acting upon them. "The true," William James said in his 1907 best-selling essay collection, "is the name of whatever proves itself to be good in the way of belief, and good, too, for definite, assignable reasons."[64] Truth, that is, was identifiable by the good consequences that it brought, given that, as Dewey put it, knowledge was inseparably connected to action.[65]

As pragmatism's many critics would point out, an ends-oriented test for judging truth was usable only when judges possessed a normative standard for evaluating the goodness or morality of outcomes, only when they possessed, to use James's quoted words, "definite, assignable reasons" for favoring the outcome. Pragmatism itself could not formulate such standards. For Leopold, however, a standard was ready at hand, and in his embrace of his standard he distanced himself from the era's pragmatists. Human life was good, human flourishing was good, and people today should keep land productive for future generations. These values were adequate to serve as a normative standard. With them, he could determine whether competing understandings of nature, once put into practice, brought good results.

A good that transcends preferences. Leopold's land ethic called for landowners and others to live in ways that sustained land health. It was thus derivative of, and a measure intended to implement, his proposed conservation goal. The land ethic, as Leopold explained, "reflects a conviction of individual responsibility for the health of the land." In his essay on the ethic Leopold summed up his goal of land health by referring to maintenance of the "integrity, stability, and beauty" of the land (biotic) community.[66] Leopold defined his terms in ways quite different from current usages, and it has been easy for readers today, using current definitions, to misunderstand Leopold's meaning. Fortunately his meanings have been made clear.[67]

By phrasing his ethic as he did Leopold distanced himself considerably and radically from ethical norms that respected and protected the individual human as autonomous being. The welfare he promoted was the welfare of the community as such, the community of which humans were a part. Humans benefited from this ethic indirectly, by their participation in the land community and the gains they got from its long-term health. Here, too, Leopold countered liberal individualism in both its conservative (pro–free enterprise) and more liberal (pro–individual flourishing) forms.

Leopold's attitude toward values and ethics was a complex one. His ethic called for humans to forge emotional ties with the land—to use it with "love and respect"—an ethical stance that seems to draw upon sentiment- and virtue-based ethical theories of Christianity (and, nearer in time to Leopold, David Hume).[68] At the same time he made clear that human survival was at stake and that the pursuit of land health was necessary for human life to continue flourishing, a claim that sounded in utility and thus resembled the views of Bentham and Mill (particularly the latter, since Leopold recognized qualitative as well as quantitative differences in alternative outcomes).[69]

The key to categorizing Leopold is to start with his view of humans embedded in the land community, a community that could be more or less healthy. The health of this community he embraced not just as the best conservation goal but as an expression of the common good. Leopold never made normative use of individual-rights rhetoric and he viewed property rights in particular as subordinate to the common good of land health. He thus seemed to adhere, in this setting, to the utilitarian view that individual rights were derivative of the common good, that is, to the view that society properly recognized and protected individual rights because and to the extent they promoted public welfare.[70] There was thus no call, in Leopold's view, to talk about a conflict between, or a balancing of, public and private interest. Put otherwise, Leopold suggested that the individual human, when interacting with nature, was understood first and best in context, as an interdependent member of the land community.

With his organic communal vision in hand it made sense for Leopold to conclude that an individual, particularly a landowner, needed to strive to fit into the natural order. Leopold's stance is easily linked to the perspective of ancient Greek thinkers for whom "one of the ultimate aims of a human life [was] to find its rightful place within the cosmic order," "to adjust and orientate ourselves to the *cosmos*," as Roman Stoics later phrased it.[71] For most ancient philosophers, nature was viewed as good and as such provided standards for human behavior that were external and superior to humankind.

> Broadly speaking, the good was what was in accord with the cosmic order, whether one willed it or not, and what was bad was what ran contrary to this order, whether one liked it or not. The essential thing was to act, situation-by-situation, moment-by-moment, in accordance with the harmonious order of things, so as to find our proper place, which each of us was assigned within the Universal.[72]

Leopold wanted nothing to do with claims of free-marketers that the market could somehow turn private selfishness into public virtue. He knew perfectly well that market-driven landowners often degraded their lands.

Bad results came, he explained in an important talk in 1939, when "everybody worried about getting his share; nobody worried about doing his bit."[73] Only with the widespread embrace of a land ethic would private land uses line up with public virtue. Only in that way would society achieve the utopian vision sketched much earlier by John Stuart Mill, Karl Marx, and others, the vision in which the interests of the individual and society came into alignment and citizens found their happiness by doing working that benefited everyone.[74] This, he believed, was the only route to a happy future time when, in Marx's terms, individuals would "realize themselves in and through the self-realization of others."[75]

At times Leopold seemed to view land health as something like an objective ideal that existed independently of human knowledge and choice. This idealist view, even during the horrors of World War II, was by no means dead.[76] Leopold grounded land health in ecological processes, embedded in nature, that transcended human choice. At other times, however, Leopold implied that land health was likely a good simply from a human perspective, from the perspective of humans and the many other species that benefited from it. Nature itself plainly did not care about particular species, humans or others; paleontology, Leopold observed in a 1941 presentation, was "a book of obsequies for defunct species."[77] In that light, land health was a human-created norm, albeit derived in important part from nature's functioning. It was thus not simply a fact about nature—not a conclusion of science alone—but instead became an ideal through conscious human choice, a combination of ecological facts and a belief that life on Earth, human life in particular, ought to flourish.

Leopold, in sum, engaged in modes of thinking that put him apart from most dominant strands of thought in his day. Taken as a whole, his package of ideas was both radical and action-oriented. He rejected liberal claims of individual autonomy and strong claims of human exceptionalism in the natural world. Though a scientist, he had doubted our ability to know and translated his humility into a call for restraint. Nature, he insisted, was best understood as a complex, organic whole—a community of life—which included humans and upon which humans in the long run depended for their flourishing. Given limits on knowledge, given the need to remedy land misuses, Leopold was willing to take action based on his best scientific guesses—perhaps suggesting an embrace of a pragmatic definition of truth, but perhaps simply realizing that the best guesses of scientists were better than continuing with clearly misguided assumptions. And then there was his normative vision, so contrary to individual ethics, a normative vision that placed value in the organic whole and linked human welfare to the flourishing, the health, of that whole.

A Radical Change

Leopold in his conservation talks tended not to go into the matter of implementation, how society might go about promoting the goal of land health, except to contend—as he did in his fourth main point—that a radical change in culture was essential. His allotted time as speaker was simply too short on most occasions to go further. He did make clear, however, that society could not rely on market forces to achieve the goal. Conservation measures sometimes benefited individual landowners but often the benefit went only to the community as such, and even then future generations needed to be part of the calculation. "Conservation, at bottom," he noted at a 1947 dinner honoring a colleague, "rests on the conviction that there are things in this world more important than dollar signs and ciphers."[78]

Given that much conservation did only benefit the community, not the private landowner, "it follows that if conservation on private lands is to be motivated solely by profit, no unified conservation is even remotely possible. Community welfare, a sense of unity in the land, and a sense of personal pride in such unity, must in some degree move the private owner, as well as the public."[79] And again:

> If cash profit be the only valid motive for decent land-use, then conservation is headed for catastrophic failure. Good land-use is a balance between utility and esthetics. It yields a highly variable mixture of individual and community profits, of cash and unponderable profits, and all accrue from investments which vary from borrowed cash on one hand to mere loving care on the other. . . . This being the case, conservation education should rest its argument on decency and social behavior, rather than on profits alone.[80]

As he commented on the failings of the market, Leopold did not blame it as an institution. He understood that it merely helped individuals as such get what they wanted. If individuals changed their wants, embracing something like his land ethic, the market would be much less problematic. He made the point in a 1942 talk:

> What we call economic laws are merely the impact of our changing wants on the land which supplies them. When that impact becomes destructive of our

own tenure in the land, as is so conspicuously the case today, then the thing to examine is the validity of the wants themselves.[81]

With frustration and occasional anger Leopold also dismissed the idea that real change could come about simply by giving people the facts so they could see how and when their activities caused harm. Evidence of harm alone, Leopold had come to see, simply did not have much effect. Leopold made the point strongly in the beginning of a writing that historian Susan Flader dates to the mid-1940s, a writing that he likely would have toned down had he continued to work on it:

> "If the public were told how much harm ensues from unwise land-use, it would mend its ways." This was once my credo, and I still think is a fairly accurate definition of what is called "conservation education."
>
> Behind this deceptively simple logic lie three unspoken but important assumptions: (1) that the public is listening, or can be made to listen; (2) that the public responds, or can be made to respond, to fear of harm; (3) that ways can be mended without any important change in the public itself. None of the three assumptions is, in my opinion, valid.[82]

At times Leopold's somber view did slip into not just public presentations but published versions of them. One occasion was his 1947 talk to a garden club, where he lamented the inability of even fellow conservationists to understand land as an interconnected community. "There is an important lesson here: the flat refusal of the average adult to learn anything new, i.e., to study," he complained. "This anger-reaction against new and unpleasant facts is of course a standard psychiatric indicator of the closed mind."[83]

Leopold openly explored ways to bring public pressure to bear on land abusers, and called for boycotts (a "pinkish word," he admitted to an audience in 1942) of the products of land coming from misused land.[84] Yet even as he talked about such organized social pressure he knew that prospects for it were dim. A government official who saw his draft questioned whether such a social-pressure approach would really work better than the measures the federal government was then using. Leopold did not challenge the claim: "I have no illusions about the workability of my plan. It will work only with people who are really in earnest, and these are few. In many fields of conservation it will not work at all."[85]

For a time Leopold also wondered whether cash-incentive programs might induce landowners to change their ways, and he briefly supported government programs that offered incentives. Leopold's hope, though, was that a landowner who by inducement began using land well would see the wisdom of good land use and voluntarily continue it when the payments stopped. Trials of the approach, though, were unpromising, as landowners tended to halt unprofitable activities the moment public support ended. Leopold viewed the trials as failures and changed his thinking accordingly. He offered his new thinking in a talk to wildlife professionals in 1939:

> I hasten to add that I no longer believe that a little "bait" for the farmer, either in cash, service, or protection, is going to move him to active custodianship of wildlife. If the wildlife cropping tradition is not in his bones, then no external force, either of my kind or any other kind, is going to put it there. It must grow from the inside, and slowly.[86]

Part of the problem was the institution of private property, which gave landowners too much freedom to use land in bad ways. "The present legal and economic structure, having been evolved on a more resistant terrain (Europe) and before the machine age, contains no suitable ready-made mechanisms for protecting the public interest in private land. It evolved at a time when the public had no interest in land except to help tame it."[87]

Private property was a useful, indeed indispensable, institution, but had to be reshaped culturally if not legally to push landowners to change their ways. "Viewing the field as a whole, we see one common denominator: regard for *community welfare* is the keystone to conservation," he told a student group. "Private land is only a stock certificate in a common biota. Private land-use must recognize an obligation to community welfare. No other motive has enough coverage to suffice."[88] The point for Leopold was of prime importance, so much so that he inserted it as the first of only two elements in his 1946 proposed conservation platform for the new political party: "The average citizen, especially the landowner, has an obligation to manage his land in the interest of the community, as well as in his own interest."[89]

What Leopold recognized, after years of failed efforts at finding more simple solutions, was that ordinary people simply had to become better than they were according to his scale of values. A radical change was required in the ways people saw the land, valued it, judged its beauty, and understood their relationship to it and with one another. A similar view has recently come from John Bellamy Foster, one of today's most acute observers of our environmental plight:

> We must reject a social system that demands the fragmentation of all living things and substitute one that promotes wholeness. If we are to save the

planet, the economics of individual greed and the social order erected upon it must give way to broader values and a new set of social arrangements, based on a sense of community with life on earth.[90]

Leopold knew that radical change did not come easily or quickly. Out of a well-conceived conservation program "may eventually emerge a land ethic," he speculated. "But the breeding of ethics is as yet beyond our powers. All science can do is to safeguard the environment in which ethical mutations might take place."[91]

At times Leopold clearly wondered whether there was any way this could come about, given the stubborn resistance of even highly educated people. As he queried in one draft: "Is the complete modern, duly equipped with a social conscience, a set of new tires, a Ph.D. in economics, and a complete ignorance of the land he came from, capable of forming a stable society?"[92] But gloom did not dominate; as he put it in a letter to colleague Bill Vogt, "that the situation is hopeless should not prevent us from doing our best."[93] Ultimately, Leopold had to rest his faith in the ability of people to learn and evolve over time, as he did in a 1939 presentation:

> Sometimes I think that ideas, like men, can become dictators. We Americans have so far escaped regimentation by our rulers, but have we escaped regimentation by our own ideas? I doubt if there exists today a more complete regimentation of the human mind than that accomplished by our self-imposed doctrine of ruthless utilitarianism. The saving grace of democracy is that we fastened this yoke on our own necks, and we can cast it off when we want to, without severing the neck. Conservation is perhaps one of the many squirmings which foreshadow this act of self-liberation.[94]

In some way, in short, people simply had to reorient themselves to land, coming, as he had, to love and respect it and to embrace other creatures as fellow community members. He summed up his conclusion in a talk to students around 1947:

> If the individual has a warm personal understanding of land, he will perceive it of his own accord that it is something more than a breadbasket. He will see land as a community of which he is only a member, albeit now the dominant one. He will see the beauty, as well as the utility, of the whole, and know the two cannot be separated. We love (and make intelligent use of) what we have learned to understand.[95]

He could only hope that his dream of a reformed humankind might one day come true.

A New Direction?

Leopold was ceaselessly impressed by the ingenuity of the modern human, particularly his cleverness in developing new products and technologies. He was equally impressed in far different ways by the seeming inability of the modern human to mature in his emotions, aesthetics, and ethical ideals. He reflected on this seeming mismatch as he concluded a talk in 1938:

> We end, I think, at what might be called the standard paradox of the twentieth century: our tools are better than we are, and grow better faster than we do. They suffice to crack the atom, to command the tides. But they do not suffice for the oldest task in human history: to live on a piece of land without spoiling it.[96]

Humans could fulfill this oldest task only if they changed their ways significantly. For that to happen they simply had to become much different people. As he went about describing the kind of person needed, Leopold ended up challenging dominant strands of Western culture since the Enlightenment, particularly the ideal of the autonomous, rights-wielding individual who could pursue his self-interest subject only to modest limits. The whole picture was wrong, Leopold proclaimed. We are fundamentally parts of something larger, plain citizens of the land community, and our first duty is to live as responsible members of that community. Reason could help us along, but we also needed much different values and aesthetic sensibilities; sounder emotions were an indispensable part. We certainly needed to cast aside the silly claim that the market could consistently turn land-use vice into healthy virtue; the evidence against that folly was simply overwhelming. Of course we needed new understandings about private property and the rights of ownership. But laws arose out of popular sentiment, so popular sentiment had to change first. Therein lay the root cause of ecological degradation. Conservation policies could succeed only insofar as they directly aimed at that root cause—only when they directly aimed at helping people become responsible, contented members of the land community.

Leopold was not unaware that democracy had its critics, that for nearly three centuries commentators from various quarters viewed it with suspicion because it lacked the power to keep ordinary wayward people in line. Leopold raised the issue in a talk when America's fortunes in World War II were dark. "Hitler's taunt that no democracy uses its land decently," he stressed, "while true of our past, must be proven untrue in the years to come."[97] But what could be done to prove it untrue?

Leopold's path was the path recommended by Socrates, Aristotle, and countless writers in between. Ideas had to be put out for public discussion and subjected to criticism. People who saw more clearly than others needed to present their arguments forcefully and push the discussion. Over time, solid facts and arguments would carry the day, or so one hoped. Socrates thought so highly of humans that he believed no one who knew what was good would fail to keep to the good;[98] the path to virtue was thus the path to enlightenment. Leopold by no means embraced this view, but he nonetheless stuck to the path of education and public discussion, hoping that in some unforeseeable way, perhaps through some invisible force of evolutionary pressure, better people would emerge.

Meanwhile and in the shadows, there was the alternative way described by Karl Marx and others. Like Leopold, Marx believed that to achieve progress in human affairs, "nothing short of a radical transformation of human nature would suffice."[99] Marx, though, disagreed with utopians of his own day who thought that opponents could be won over purely through argument.[100] Society was not a battle of ideas, whether based in fact or on moral grounds. Economic forces and practices were in charge; the modes of production were what formed the base upon which all else—ways of thinking, social organizations, and more—arose. Without fundamental change in the economic system, in the modes of production, the existing order would remain and prevailing ideas and values would persist. Like Marx, Leopold recognized the need for significant changes in economic practices, yet he believed—or assumed, or perhaps merely hoped—that ideas and new moral visions could help bring change about, without need of class conflict and revolution.

Reading Leopold today one cannot help but wonder what might have happened had the conservation movement of Leopold's day, and since, listened to his last talk and taken heed. As events unfolded, though, conservation continued on the trajectory that Leopold criticized, attending to the specifics of land- and resource-use practices and, around 1960, taking on a stronger concern with pollution and contamination. The movement remained fragmented with groups working at cross purposes. It never took on anything like an overall goal, Leopold's or any other. The movement did not identify bad culture as the root of the problem, and made no real effort to change the ways people saw the land and their place in it. It did little to question the dominance of individual liberalism and autonomy and offered no conservation version of what private landownership ought to mean.

Today, as in Leopold's day, people do not see the land as a community of which they are a part; they do not realize that this community can be more or less healthy; they do not see land health or anything like it as an overarching conservation goal and indispensable to long-term human welfare. Leopold has been much cited while his main messages have not been taken to heart.

With the environmental cause stumbling—most vividly on the issue of climate change—perhaps the time has come to take a more radical stance in the way Leopold proposed, a culturally radical stance that forcefully introduces new ways of thinking and valuing, new ways of understanding the human predicament, new ways of talking about burdens of proof and what qualifies as truth, new ways of identifying the human good. Leopold believed that only radical cultural change could lead to a healthy future. The years since then have not shown otherwise.

TWO

The Love of Wendell Berry

Now, we got to get this thing right. What is needed is a realization that power without love is reckless and abusive, and that love without power is sentimental and anemic.

—Dr. Martin Luther King Jr.

I am aware that invoking personal decency, personal humility, as the solution to a vast risk taken on our behalf by corporate industrialism is not going to suit everybody. Some will find it an insult to their sense of proportion, others to their sense of drama. I am offended by it myself, and I wish I could do better.

—Wendell Berry

A central theme in the writings of Wendell Berry—maybe the most important of them—is his concern about relationships of many types and about the practical and moral urgency of tending them. The world that Berry observes, looking out from his hillside, northern Kentucky farm, is not made up of physical parts in isolation. It is not a world of individual people, distinct tracts of land, and particular natural resources. It is composed instead of interconnected elements, and the links and bonds among them, the interdependencies, are as real and vital as the elements themselves. We have neglected these many connections, Berry tells us pointedly. We too readily see the world in fragmented terms, valuing its parts in isolation and ignoring or undervaluing the ties that ought to bind. This fragmentation in perceptions and standards extends beyond the physical realm. In intellectual and moral realms we are also prone to fragment integrated wholes into parts, weakening connections and diminishing the good that comes when wholes are kept intact and flourish.[1]

Berry's embrace of an integrated, holistic worldview situates him where he can roundly criticize much of contemporary society. It is little wonder that his extensive writings—novels, essays, poetry, and short stories, a good five dozen volumes—range so widely, from soil erosion, to abortion, to free trade, to the curricula of universities. Fragmentation bears ill fruit in all of these settings and more, Berry observes. The solution to weak or torn relationships is to recognize their value and somehow mend them, by respecting ecological processes, for instance; by becoming good neighbors and community members; by connecting secular and sacred; and by linking the present generation with past and future ones. It is a message, a call for renewed and expanded virtue, that resonates with Berry's numerous readers, as well it should. It helps account for his stance in the eyes of many as the preeminent voice of conservation of his generation.

Environmental decline, like decline in social communities, is plainly linked to an excessive individualism in social realms and to widespread tendencies, when dealing with nature, to fragment and commodify its pieces and parts. As Berry recommends, people everywhere should slow down, sink roots in place, join hands with neighbors, and act faithfully toward local lands. His is a call for virtue and attentiveness to good character, which is to say for reform at the individual level, even as he holds high the community as such. He offers advice he has long followed himself, ever since he returned a half century ago to Kentucky with the aim "to make myself responsibly at home in my native and chosen place." His has been a "long term desire," he has explained, "proposing the work not of a lifetime but of generations."[2]

Berry develops this moral, communitarian theme perhaps most fully in his novels and short stories, all set in a fictionalized version of the hilly lands around his home along the Kentucky River. As he recreates this world in his fiction, Berry includes many characters who draw the reader's admiration because of their good traits and their attentiveness to communal roles. No character, of course, is perfect, but a few get close to it. In them we gain a sense of the kind of people we ought to become. We get a sense of how, person by person, we might move some small part of the world—a particular place on Earth—in a better direction.

Berry's focus on individual virtue inevitably raises recurring questions of long standing. The forces pushing down on his home landscape and places like it, urban as well as rural, largely come from the outside, the cultural forces as well as the market-driven economic ones. This he well knows. In light of these external pressures, though, how much can an individual accomplish by acting virtuously, other than to minimize involvement in the forces that foster decline and to set good examples for others? Both in nature and in social realms, the welfare of the parts often depends upon the welfare of the integrated whole. The health of that whole, in turn, can depend upon coordinated corrective action undertaken at the level of the whole, by people acting together both as land users and as citizens. Individual reform—in values, perceptions, even daily conduct—no doubt can take us partway down the path of change, partway to a place of hope realized. But how far can it go? Further, to lay stress simply on change at the individual level, without more, poses some significant dangers. It can risk reducing complex economic and social problems to matters merely of private morality, in the process diverting attention from insistent needs for structural change. It can end up blaming individuals for the ill consequences of powerful forces that surround them and sweep them along without adequately attending to the institutional structures of society in which individuals necessarily function.

Berry's overall vision, once plumbed and assessed, falls short of a complete strategy for moving ahead and it is helpful to see why that might be so. Nonetheless, his guiding moral vision, his penetrating prophecy, is decidedly vital if not essential. Berry stands as perhaps our sturdiest and most far-sighted land-based cultural critic, laying bare many of the root, cultural sources of illness in lands and people. We have failed to see those root causes clearly, instead mostly attending to their superficial manifestation. Berry helps us confront them.

Environmental reform in all guises needs to be guided by visions of wholeness, social, natural, intellectual, and moral. Similarly, reform efforts need to come together, both individual reform and collective action, to give rise to something larger than its parts: to a cultural push for the resettlement of the continent; to a new understanding of the human place in the community of life; and ultimately to new structures of daily life that help people as individuals and families live the kinds of moral, satisfying lives that Berry so warmly presents.

Relationships

Wendell Berry is a married man in many ways, and he wants other men, women, and children to marry as well. He is bound to his wife and to his children and grandchildren. He is also bound to his land, to his community, and to the many generations that lie dead on the hill above his riverside farm. Though as a farmer he uses livestock and kills them in time, the human-animal bond on his side reflects respect and honor. We live with animals, Berry says, as much today as in the past. They are hardly our equals, but they are morally worthy creatures in their own particular ways.

These many paths of interconnection extend to memories, loyalties, affections, and cultural traditions. To live well in a place, Berry relates, is to become part of this encompassing community of life, entering into its memories, reworking them, and then passing them along. This type of cultural interdependence is just as vital as the cycle of natural fertility, which begins (in the case of terrestrial life) with the soil, moves through plants and animals, and returns in time to the earth by death and decay, making way for new life. It forms overall a great wheel, Berry tells us, linking present, past, and future life in a ceaseless flow. For all life to flourish, these many cycles and links must remain intact. As did English reformer Sir Alfred Howard, Berry views "the whole problem of health in soil, plant, and animal, and man as one great subject."[3] For him as for Aldo Leopold, the overall health of the land is "the one value," the one "absolute good."[4] And it is the community that is the smallest unit that might rightly be called healthy. "To speak of the health of an isolated individual," given the individual's links to the whole, "is a contradiction in terms."[5]

In thinking about the necessary requirements to achieve land health in farm landscapes such as his own, Berry has returned again and again to questions of economics. Good land use must be economically feasible, it must support the farm family and cover costs with sufficient security and margins to give owners a sense of long-term security. Farm owners need confidence that they can pass their lands down to the next generation, which can similarly flourish. Just so, land health depends on the existence of a supporting local community of land tenders who develop, remember, share, and augment local wisdom on how best to live in a particular place. The well-being of any single farm is thus linked to the well-being of surrounding ones. The welfare of the present generation is linked to the plight of earlier and later ones.

Berry's commitment to relationships shows up in nearly all of his cultural criticism. His concerns about abortion, for instance, challenge the ways abortion illustrates and accepts weaknesses in marital bonds and in attitudes toward child-bearing. Berry worries about feminism for the same reason: not because it honors women but because it threatens to value the human parts in isolation from the roles that women (and men) fulfill. It can too easily portray particular women apart from the roles in families, neighborhoods, churches, and work that yield meaning to their lives. As moral beings men and women possess equal value, but the many tasks and roles of daily living are varied and call for different skills. Prior generations faced these challenges and crafted time-tested ways of addressing them. We should be slow to set aside that wisdom, slow to criticize traditions simply because they collide with equality ideals abstracted from context, group needs, and practical economics.

Higher education similarly incurs Berry's criticism because too often it treats students as isolated creatures, not as situated interdependent beings. It trains them to become cogs in an economic system that views labor and nature as commodities or inputs while ignoring the duties and satisfactions of belonging to place. To study science apart from philosophy, literature, or history is to deny the many bonds that link human experiences and knowledge into an integrated, evolving whole. To distinguish book learning from practical experience—and both from the wisdom of elders—is to presume a division if not a hierarchy of learning that can lead people astray. Work and leisure should flow together, Berry tells us, and so should the secular and religious. Wholeness arises when labor overlaps with prayer. The good life is one in which roles are not distinctly separated and in which a person's good character is never set to one side.

Berry illustrates these connections in a brief essay, "Contempt for Small Places," in which he expresses hope that people someday will see the links between the Gulf of Mexico's "dead zone" and strip mining taking place in Kentucky. In this instance, water provides the connecting link:

> The health of the oceans depends on the health of rivers; the health of rivers depends on the health of small streams; the health of small streams depends on the health of their watersheds. The health of the water is exactly the same as the health of the land; the health of small places is exactly the same as the health of large places. As we know, disease is hard to confine. Because natural law is in force everywhere, infections move.[6]

A more developed expression of the theme appears in Berry's masterful short story "The Boundary." In it, an elderly, tottering Mat Feltner inspects the wire fence that surrounds his hilly farm to see that it remains taut enough to contain livestock. At first glance the story appears to exalt the sturdy fence, much as does Robert Frost's poem "Mending Wall." Certainly the elder Feltner, a character whom Berry clearly admires, is insistent on keeping the fence strong. Yet, even more than does Frost, Berry pushes readers to question fences and to see the dangers of them.

Mat Feltner's fictional boundary contains his cattle but little more than that, physically or in Mat's mind. The fence permits life to flow over and through it. Wild animals also pass the boundary with ease. As landowner, Mat is linked to his neighbors in ways that generate benefits that transcend the legal lines. And as the narrative unfolds and Mat's warm memories take over he displays to us just how connected he is also to past generations, to his extended family, to the community itself, to future land tenants, and most vividly to his elderly, worried wife, waiting for him in the house on the hill. Mat's thoughts on his farm's productivity are not detached from his thinking about its beauty; his utilitarian calculations of market profit are not cut off from the farm's aesthetic and spiritual harvests. Mat accepts responsibility for his individual actions, yet he entertains no illusion that he can go it alone. Particularly on ecologically challenging land, good farming necessarily builds upon local, experience-based wisdom that takes generations to arise. A handing down is thus needed, from generation to generation, good owner to good owner. As for the behavioral misfits that afflict all families and places, good ties are essential here as well. Thus, the survival of the alcoholic Uncle Peach (in the story "Thicker than Liquor") depends on the willingness of family members to recognize their family duties and to step into them.

Perhaps inevitably Berry is drawn to circular images when he talks about life and its unfolding. Endings become new beginnings, or they need to if life is to flourish. And so we recognize fragmentation in its many forms as the bane of fertility and health, for human communities and natural ones alike. Glad and Clara Pettit promote fragmentation (in the story "It Wasn't Me") when they deny their bonds to Clara's recently deceased father, Old Jack Beecham, and to the farm tradition of which he was a part. For the city-dwelling Pettits, the rural Beecham farm that they inherit is simply a capital asset, properly valued for its annual crop yield—which is to say valued apart from its functional roles in the surrounding natural and human landscapes. It is left to the family lawyer Wheeler Catlett and good community leaders like him to strive to keep these vital local bonds intact—in the instance of Jack Beecham's farm, to promote an orderly handing down of the farm from Jack to the current tenants, Elton and Mary Penn, and beyond them to the good tenders next in line. Jack's scribbled instructions about his last wishes for the farm (selling to the young Penn at below market value) are construed by Catlett in the context of Jack's life and of his agrarian culture; they are interpreted, that is, in a way that honors the connections between Jack's words, his life, his good tenants, and the needs of the land. The lawyer Catlett urges the Pettits to honor Old Jack's community-supporting wishes, even though his scribbled notes are not legally binding. The small-minded Pettits, though, see matters much differently. Out of context, seen from the city, Old Jack's scribbled words make little sense. For the Pettits, the only words that matter are ones the law will enforce.

Standing Apart

One hardly needs to dip into any of Berry's writings, in any of his genres, to recognize his detachment from much of modern culture. The contrast is especially stark when we compare his work with realms of modernity in which fragmentation has proceeded furthest, with aspects of industry, the market, and education. Industry and the market view nature as warehouses of discrete commodities, to extract, process, buy, and sell. Parts are judged and valued in isolation, not in integration. Interconnections as such count for little, and future costs and benefits are heavily discounted.

We can see this dominant mentality at work in the economic treatment of the individual tree. As it grows and interacts with its surroundings, year by year, it adds no value to national income when economists go about calculating the periodic GDP. Value arises only when the tree is cut and hauled away. Then the tree's value is assessed in market terms, as lumber, in terms that ignore its place of origin. Meanwhile, the loss of the living tree is economically irrelevant—it is not counted as a cost—as are the ecological effects of harvesting it. Contrasting with this attitude is the perspective of the caring landowner, who values the tree as it grows and appraises it as part of a forest or woodlot, as well as alone. The caring owner knows, too, that losses come when a tree is removed and that it is prudent to keep losses modest and to replant promptly.

What is true of the tree is true of nature's other components. A flow of water that a person deploys for irrigation has commodity value only when it is separated from its source. (In the western United States, a prior-appropriation water right only arises if and when a person does divert the water.) Water has no market value when left in place to sustain the river. The river as a whole has no value given that it cannot be bought or sold. Tracts of farmland valued for yield are assigned prices with little or no regard for location or surroundings. Human labor is treated much the same, mostly valued at the going commodity rate.

As for the academy, intellectual fragmentation is greater than outsiders usually realize. Work that one department values, given its peculiar culture, another department might reject as nearly worthless. Critics have long lamented the separation of science and literature, yet the fragmentation continues apace, within fields of study as well as between and among them. In the case of farmland, industrial scientists are happy to manipulate nature at will, down to the smallest genetic unit, paying only modest regard for the effects of farming on ecological processes, landscape beauty, and wild-life populations—natural elements that researchers elsewhere in the academy value highly. While scholars in one corner probe the limits on human knowledge, scholars in another charge ahead as if knowledge were complete. While one group ponders the extension of moral value to other life forms, another group remains confident that humans alone possess worth. While one group, committed to mathematical models, touts the ability of the market to move nature's parts to their highest uses, another group, studying actual places and activities, catalogs the many ways that market processes bring degradation.

More and more, academic research by definition entails the collection and analysis of factual data in ever-more specialized niches. Meanwhile, few people think seriously about the whole of things, landscapes included. What is good land use, for instance? And what kind of culture is needed to promote it? The questions are rarely phrased because they fit within no specialty. Academic fields charge ahead with proposals to reshape the land, without any comprehensive way to evaluate the results. Is it any wonder that the works of Berry himself are pushed to the academic fringe? Is it a surprise that few academics can imagine useful truth embedded in fiction or poetry?

Fragmentation in American culture, Berry illustrates, is particularly evident in social and political thought. Step by step we have "liberated" individuals from social shackles, which is to say exalted people as autonomous individuals while minimizing the centrality of loyalties, duties, and interdependencies. Countertrends do exist, to be sure—the billboards encouraging men to become good fathers and families to take in foster children. But even here we must ask whether the parent-child relationship has independent or collective value or whether it is, instead, more an instrumental means to individual ends. America is the paradigm land of liberalism in the classic sense that exalts individual choice. Though often labeled conservative by the political right, the economic libertarianism of Wall Street stands squarely in America's tradition of liberal autonomy. So does the social liberalism of what is termed the Far Left. Time and again, we reduce social disputes to competing claims of individual rights—the right to an abortion, to own guns, to drain wetlands—as if the activities being defended did not implicate neighbors, family members, and the community as a whole. For the economic libertarian as well as the ardent left-liberal, relationships are mostly voluntary and transient, lasting only so long as an individual wants them to last. At the outer edge we find the liberals known as contractarians, who challenge nearly all responsibilities save those agreed to voluntarily. Responsibilities are few and malleable.

Berry's dissent from this kind of fragmentation appears vividly when we turn to food and to other basics of life. Berry seeks to connect eaters to food producers and to specific food sources on the land; he is the leading light of the local food movement. The free market, in contrast, pushes in the opposite direction, reducing food transactions to matters of price and cash flow. Berry encourages consumers to engage morally with the methods used to produce their goods (the pollution, land degradation, social displacement, child labor) and with the fates of their post-consumption wastes (the leaking landfills and disrupted fertility cycles). Here again, he stands apart from the dominant market view in which packaged goods appear on store shelves shorn of histories and thus moral complications. The same shallow cash-nexus weakly connects the consumer to the modern waste hauler and landfill operator.

In Berry's view (borrowed from agrarians of the past), barter rises above the cash sale in supporting a sound relationship between bargaining parties. Best of all are links among friends and neighbors involving exchanges of labor that are left unrecorded. Not cash and bargaining, Berry tells us, but love itself forges the most durable bond. Only love can bind all, weak though it sometimes may be.

Theories of Progress

Implicit in much of Berry's work are claims that we can move ahead as a people, or at least recover lost ground, by acting virtuously and improving our many relationships. To the extent Berry offers a theory of progress—or a hope or vision of progress—this is likely it. To understand this component of his thought, putting his contributions in context, we might usefully compare Berry's views with three competing perspectives on human nature and progress. Ultimately environmental reform needs to have traction in some way. It needs to bring about new ways of living if not thinking and valuing, countering the ill winds of the modern day. How might this come about?

Jacksonian democracy. Growing up in rural Kentucky Berry naturally embraced elements of the political ideology long known as Jacksonian democracy: a commitment to individual initiative, free enterprise, minimal government, and a safety net provided by family, friends, or parish rather than society. Like other lovers of liberty, Berry recoils from vestiges of feudalism and binding hierarchy (slavery most of all). In his various writings he embraces not just free labor, but something akin to a labor theory of value, particularly for skilled craftsmen. Antebellum democrats held high the ideal of the independent farmer or small businessman, beholden to no baron or bank. Especially in his fiction, Berry still exalts the ideal.

Where Berry stands apart from this still-lively strand of political and economic libertarianism is in his rejection of the Jacksonian theory of progress. Jacksonians assumed that a release of entrepreneurial energy would somehow lead automatically to improvements in the public good. A popular explanation employed Adam Smith's invisible hand: as individuals aggressively pursue their self-interest, the theory claims, the market magically transforms their selfishness into public virtue. Private vice could yield widespread gain. Few Jacksonians may have read Smith or understood his logic. But the spirit of negative liberty was alive on the land.[7] Antebellum Americans wanted both liberty and progress, and felt sure that they would travel hand in hand.

Berry largely rejects this belief that the market can transform individual vice into shared good. We see his resistance clearly in his fiction, where good farmers such as Mat Feltner, Nathan Coulter, and Elton Penn think constantly about their local community and its future as well as about themselves and their immediate families. These leaders stand apart from characters such as Troy Chatham in the novel *Jayber Crow*, who is quick to borrow, expand, and add equipment in a headlong pursuit of wealth. When the deceased Jack Beecham's farm is up for auction, we have no illusion that the best new owner will be the one who can pay the most for it. When Troy Chatham sells for top dollar the Keith family's "Nest Egg"—its ancient wooded tract, now valuable for timber—his foolishness is evident.

Berry's rejection of Adam Smith, though, seems to leave a void in his own embrace of responsible individualism. Selfishness is not soon to disappear and the good conduct of the best people will not be mimicked by all. If the market is not strong enough to turn selfishness into public progress, how then can it be contained? Without Smith's invisible hand, how can we respect the individual liberty of landowners and others and still end up with ecologically healthy lands? The question has loomed long and large. Berry is hardly the first to confront it.

Marxism. Berry's writing also contains striking parallels, little noted, with the fundamentals of Karl Marx's thought and a comparison of the two sheds good light. The overlap is most evident with Marx's critique of capitalism and "bourgeois" property as well as his emphasis on the social and cultural implications of a society's chief modes of economic production. As in the case of Jacksonian democracy, we find with this comparison commonalty on key elements alongside a disagreement on how and why progress might take place. In brief, Berry shares most of Marx's criticisms of market capitalism, yet rejects Marx's ideas about the means and processes of long-term change.[8] Unlike Marx and those following in his tradition, Berry has little confidence that unseen forces, economic or otherwise, will likely bring about the needed change: Ordinary people must do it themselves.

Writing mostly as an economic historian, Karl Marx put forth an extended critique of industrial capitalism. In simple terms, Marx anticipated that class conflict in time would increase sharply and lead workers (the proletariat) to displace capitalists and end worker exploitation. The ultimate outcome, Marx hoped and sometimes expected, would be a peaceful, classless society in which the needs of all were met and conflict ended. Government for Marx was simply a tool of the ruling class and would fade away as classes disappeared. Marx spoke only vaguely about his ultimate vision of classless harmony. What interested him more were the manifest ills of his days, of the industrial order, and the likely first steps of the journey forward. Marx railed against the cash nexus and "the icy water of egotistical calculation," which degraded loyalties and interpersonal bonds.[9] Industrialization sapped the craftsman of all sense of pride, he complained. It transformed even professionals into paid wage laborers. All labor was demeaned as the economically powerful dominated the powerless.

As for private property, Marx supported it when used by an owner personally (as habitation, for instance, or as an owner-run trade or business; he was by no means opposed to all private property). What he attacked was property in the form of capital; that is, productive plant, equipment or land that an owner used by hiring the labor of others, paying a set wage (as low as possible), and then controlling the overall output. In this ubiquitous arrangement, the laborer lost control of his work as well as control over the value created by that work. He was alienated from his labor, as Marx described it. It was thus the factory owner and the farm landlord who drew Marx's wrath, not the independent farmer or shop owner, or the employee-owned business.

Marx's popular appeal in his time largely rested not on his hard-to-parse meta-history but instead on his embrace of the popular labor theory of value (producerism, it was often termed), a theory he wrested from economic liberals and transformed from a defense of private property (as it was under John Locke) into a weapon to attack capitalist accumulation. Workers should own and control the fruit of their labors, that was the moral claim. Their produce should not be confiscated or otherwise controlled by capitalist owners. Similarly, workers should own and control the tools they used in their work, and decide how and when to labor. Only by doing so could they retain their pride. By subverting worker control, bourgeois capitalism turned people into mere machines. Eventually, Marx predicted (wrongly, so far), capitalism would drive wages down to subsistence levels with profits going entirely to capitalist owners. In time, though, industrial workers would awaken to their plight, to their bondage, or so Marx hoped. Conflict would ensue. Capitalism would fall as society ascended to a higher cultural level and economic democracy took hold.

As historian Eugene Genovese has recorded, probably the most direct, home-grown analogy to Marx's critique of industrial capitalism in the United States has been the body of thought known as southern agrarianism or conservatism that Berry carries forward. (Once a Marxist himself, Genovese became, like Berry, a proponent of ideas selectively drawn from the South's conservative heritage.[10]) Much as Marx did, Berry stresses the central importance to society of the means used to produce basic commodities. A mode of production based on small farms and independent craftsmen, Berry suggests, leads to a moral order quite different from the one that results when production comes instead from industrial farms, factories, and global enterprises. Echoing Marx (and, even more, Proudhon) Berry calls for mutualist, cooperative enterprises that displace ruthless competition. Berry, of course, does not use the language of class, yet his comments about community bear distinct resemblance to class-based rhetoric. All people who dwell in a place belong to the resident community, whether they know it or not, Berry contends. We hear an echo in this of Marx's "false consciousness," in comments (for instance, by fictional Burley Coulter and others) about the inability of many community members in Berry's fictional small world to awaken to their true memberships.[11] And we see faintly, too, a vision in Berry's work of a small-town classless society in which social hierarchy disappears, solidarity prevails, and people equate their personal welfare with the welfare of the whole. Berry does not write about a world without government, yet government plays almost no role in the fictional world he creates.

Even more than in the case of Jacksonian democracy, though, we need to avoid pushing this comparison too far. Far more than does the materialist Marx, Berry sees power in ideas and values and he holds out the possibility of real moral progress. (Marx scoffed at the notion.) Berry pleads with readers to consider a change of heart of a type that Marx would have rejected. Indeed, Berry's individualism distinctly distances him from Marx, who challenged liberalism and saw little prospect that individuals as such could improve on their own volition.

More pertinent here perhaps is the lack of overlap in terms of Marx's theory of historical change and how it comes about. A strong believer in the determining force of history, Marx predicted that class conflict would likely arise and usher in a new era: just as capitalists as a class once displaced the feudal landowners, moving society from feudalism to the industrial age, so too workers would awaken to their inherent solidarity and overcome their capitalist exploiters. On his side, Berry makes no such sweeping claims. Though a Christian and hopeful that God's kingdom will one day come to Earth, Berry makes no mention of inexorable historical forces that will move the world ahead. He does not await a Second Coming.

When Marx's engine of change is put aside, as Berry has done, his theory of progress is cut away. Progress then has no way to come about. The ills of industrial capitalism presumably would linger on.

Civic republicanism. A third point of contrast for Berry's perspective, particularly useful on the issue of progressive change, is offered by civic republicanism, a body of thought in the Anglo-American world that gained strength in eighteenth-century Britain, which greatly influenced colonial thought just before and during the Revolution, and which survived in diminishing form thereafter. Republicanism remains alive in various strands of communitarian thought and in older, largely displaced forms of political conservatism. In academic writing it experienced a resurgence in the 1980s and 1990s.

Civic republicans (Thomas Jefferson among them) worried about the corrupting influence of the market and the aggressive pursuit of self-interest. They worried also about the loss of independence that came when one person became subject to the economic or social control of another—as an urban laborer, for instance, or a farm tenant. Good governance could emerge, they believed, only if talented citizens gained economic independence, rose above self-interest, and then engaged with one another in the important work of collective self-rule. Virtue was the key to it all, and governments were best designed when they promoted that virtue. Private interest and the public good quite often diverged. Unless checked, the selfish ego threatened civic stability, particularly when private interests manipulated government to their advantage.

In this third body of thought we again see similarities and differences with Berry's views. Like the civic republicans, Berry perceives a clash between the common good and the aggressive pursuit of self-interest. He agrees, too, that leadership requires virtue and that virtue is endangered, particularly by money. Good governance, in turn, is a communal aspiration, and people who provide good leadership deserve praise for their service. The highest calling, for Berry as for civic republicans, is for a person to rise up to become a community's exemplar, leading by individual virtue as well as by community-minded acts. In civic republican thought, community leadership meant involvement in collective government. It meant deliberating with fellow citizens about the common good and how they might, together, promote it. But as political scientist Michael Sandel explains,

> to deliberate well about the common good requires more than the capacity to choose one's ends and to respect others' rights to do the same. It requires a knowledge of public affairs and also a sense of belonging, a concern for the whole, a moral bond with the community whose fate is at state. To share in self-rule therefore requires that citizens possess, or come to acquire, certain civic virtues.[12]

It was this vision of self-rule based on virtue and collective action that faded in the decades after the 1790s. Liberty in the republican tradition was about joining together to rule in concert for the common good. The new version of liberty, embedded in economic liberalism, was more individual and negative. It depended "not on our capacity as citizens to share in shaping the forces that govern our collective destiny but rather on our capacity as persons to choose our values and ends for ourselves."[13] The shift seemed liberating and empowering, but it remained so only when and so long as individuals as such could control their destinies and so long as their choices support human and natural flourishing.

Civic republicanism is particularly useful in clarifying Berry's relative inattention to any theory of progressive change. Where (as noted) Jacksonian democrats placed faith in the market and Karl Marx deferred to the forces of history and class conflict, civic republicans have looked to virtuous leaders to craft policies to promote the common good. They have endorsed structures of government that draw in qualified leaders, diffuse power, and diminish the chance that selfishness can corrode public welfare. Ultimately, progress here rests on concerted action, on what America's revolutionaries termed their "self-rule" (a revealing label, given our tendency now to equate self-rule with individual liberty). Republican progress came when good people set aside their personal economic interests and through shared self-government labored in concert to promote the common good. To the extent Berry presents a theory of progress, it seems to take off from this civic republican base. The call to virtue, to community spirit, is certainly present. But what about the call to step into civil leadership?

The Responsible Individual

On balance, Berry leaves us with the suggestion that social progress will come, if at all, when people become more aware of their relationships and then labor to improve them. Progress, that is, requires people to improve themselves by becoming better members of families, communities, and home landscapes. The market's invisible hand will not bring about improvement, particularly to the land, Berry implies. Nor can we await some inexorable force of history to bring healing. People need to act, one by one, by becoming more moral.

Berry on occasion has expressed this message directly. "Our damages to watersheds and ecosystems," he tells us, "will have to be corrected one farm, one forest, one acre at a time."[14] Similarly, "the real work of planet-saving will be small, humble, and humbling. . . . Its jobs will be too many to count, too many to report, too many to be publicly noticed or rewarded."[15] Much of the needed change must take place within us as we cut ourselves off from the diseases that permeate modernity. "We, each of use severally, can remove our minds from the corporate ignorance and arrogance that is leading the world to destruction."[16]

Berry's emphasis on individual initiative shows up also in his criticism of social "movements." He is not out to found a movement, he tells us, and does not want to be part of one. Just as telling, his dozens of books contain scarcely any recognition that the decades-long labors of environmental groups have brought any good, to Kentucky or any other place; no recognition, for instance, that the federal Clean Water Act—the product of collective conservation efforts—has greatly improved water quality in his native Kentucky River. Indeed, in essay after essay we hear hardly a word about organizing, joining groups, or getting together with fellow citizens to push for political change. Berry promotes the idea of community-supported agriculture and organized sustainable farming and forestry initiatives. But his rare mention of such local efforts highlights his relative silence on the subject overall.

We see this limit clearly in Berry's masterful novel *Jayber Crow*, in which town barber Crow, a long-time bachelor, struggles faithfully to respond to the ills of modernity, particularly the arrival in his small-town world of the market mentality and industrial farming. Berry as novelist is true to his fictional setting when he has Crow and his fellow good citizens stand back and do nothing while industrial capitalism drags down their town. His best option, Crow senses, is to respond individually as a good Christian and love his neighbors, all of them. It is sound moral advice so far as it goes—and extraordinarily hard to embrace, as Berry vividly displays. Yet we can hardly be so naive as to think that Crow's Christian love will halt the industrial juggernaut. In time, Port William's lands will pass into the hands of farmers willing to push them hard to make money. When one aggressive farmer falls, another will take his place. Berry, to be sure, is not the fictional Crow, nor is Crow put forth as a model citizen. Yet the reticence we see in Crow runs through much of Berry's writing.

We can say that Jayber Crow acts irresponsibly in his reaction to industrial agriculture, and perhaps he does, if he thinks his love will somehow soon improve the land. We could complain, too, that Crow is more interested in assuaging his conscience than promoting the common good, though we need to go easy here because the moral life is itself a worthy and difficult end. The important point, though, is that we find little evidence, in this novel or elsewhere in Berry's volumes, of any call for concerted action by citizens acting as such—for self-rule of the kind that America's founders sought when they broke ties with Britain. We can admire the virtuous, sensitive individuals like Jayber Crow while at the same time complain, pointedly, that little is being done to halt the rural decline; little is being done to translate robust conceptions of virtue and flourishing into institutional and policy changes.

Perhaps the only exception in Berry's fiction to this disinterest in collective politics is provided by Wheeler Catlett, a lawyer who regularly features in Berry's writings. Catlett actively supports the Burley Tobacco Growers Cooperative, which endorses tobacco-production limits to keep tobacco prices high and farm incomes up. The example, though, stands out for its rarity in Berry's dozens of books. It is telling, too, that Catlett's tobacco group (like its real-world version) promoted the economic interests of its members, much like countless other industry lobbies and—even worse—promoted a product distinctly harmful to human health. Burley's group was not a civic organization that benefited supporters only as community members.

Here and there, Berry emphasizes key ways in which structural change is urgently needed: Governments ought to protect small towns from invasion by the market as well as by foreign troops, he tells us. Global free trade is too often a destructive force, and corporate powers have exceeded all reasonable bounds. Public agricultural policies are also miscast, Berry stresses, and regulatory agencies that are supposed to hold strip-miners accountable too often defer to their wishes. Even in the case of private property, which Berry supports, he sees need for fundamental legal change, halting land uses that entail obvious degradation. But one wonders, how are these many structural changes to take place if good people remain silent? How are they to come about if no citizen movement pushes hard for them?

Ultimately, what we see in Berry's work—and he presents it as finely as any author—are images of the responsible individual, the farmer or barber or tradesman who stands tall on his land or doorstep, lives as virtuously as possible, cares for neighbors, respects other life, and lends support for the surrounding community. They are good and indispensable images, valuable insofar as they can take us. But they do not take us all the way. And by dropping us off where they do—well short of the goal—we are left vulnerable. By exalting the responsible individual and calling for individual reform Berry implies that (a) individuals are mostly to blame for our ills; (b) individuals can change their ways if they would only do so (albeit the labor might prove hard); and (c) overall reform can come one convert at a time. There are problems, though, with these claims, sizeable ones, and their inadequacies pose dangers.

The Populist Persuasion

In his work Berry openly borrows from the writings of earlier agrarians, particularly the authors of the 1930 manifesto *I'll Take My Stand*. Less clear to many readers are the similarities between Berry's work and the stances of farm- and labor-based reform efforts in earlier American eras. In his call for wide-ranging popular change, beginning at the bottom, Berry's advocacy resembles messages of the Patrons of Husbandry (the Grange), the Farmers' Alliance, the Knights of Labor, and other nineteenth-century efforts that led to the Populist movement of the 1890s. These reform groups saw themselves pitted in conflict with the new industrial order. They were guided by memories of earlier days, when farmers and craftsmen had greater control over their plights. In the view of Gilded Age reformers, the era pitted people against the vested interests. Proposals for economic reform were blended with calls for social and moral change—including, in the case of the Farmers' Alliance, vast improvements in the public status of women.

For the most part these postbellum reform efforts produced little, except in the regulation of grain elevators and railroad rates. With the rise and (in 1897) rapid fall of the Populist Party, reform efforts shifted gears—or, more aptly, one reform template gave way to another. The Populists' inclusive, wide-ranging calls for moral reform and economic and cultural change gave way to more narrowly focused efforts to achieve limited results, often economic ones. In the realm of labor, for instance, the inclusive vision of the Knights of Labor yielded to the AFL and its smaller craft unions, which sought to wrestle economic gains for their skilled members, not to change society. Farm-related reform efforts divided into numerous groups, again with narrower, targeted aims, leading in time to the rise of the group that worked aggressively not for all farmers but only for the large, industrial few—the American Farm Bureau Federation. In effect, the Populist impulse provided the background for the varied reform efforts that composed the early twentieth-century Progressive movement.

Efforts by many of these new Progressive reformers did in time bear fruit. And they did so because leaders learned from Populist failings. Moral outrage, leaders realized, was not enough to bring about change, even when directed against specifically described ills and when promoted as mass political action. Better organization was required, leading to real political and economic power. "It became clear" to the new reformers, historian Maury Klein recounts in *The Flowering of the Third America*, "that power could be obtained only through organization, and that those who organized effectively did far better than those who did not. The tighter the organization, the narrower its aims, and the greater the resources at its command, the more likely it was to achieve its aims."[17] Individualism remained an icon and object of folklore, but success at law and in the market required more orchestrated, focused efforts. Craft unions gained ground when they accepted the existence of industrial capitalism and negotiated on bread-and-butter basics. Advocates of various reform efforts learned the same lesson: organize, keep focused, and push hard. Independent individuals were pawns of the system. And to succeed, a political party, a mass people's effort, had to do more than express moral outrage.

The Progressive experience that followed Populism also yielded other vital lessons. Voters, it seemed, were inclined to constrain economic enterprises only when the evidence of public harm was fairly obvious. In addition, the public's attention span on reform issues was typically short. Reformers had only a brief window of opportunity to achieve gains. To endure, real change had to become part of the bureaucratic and organizational structures of society; otherwise, gains could fade away. Even then, corporations were displaying a disturbing ability to mute calls for change by modifying their activities slightly and presenting an appearance of reform. And once the public's attention faded, corporations stepped in to takeover regulatory agencies, weakening reform efforts and bending regulations to their advantage.

Berry's writing presents instructive parallels with the efforts of early Populist reformers, especially the Grange and Knights of Labor. The parallels are hardly exact, but they are apt in that Berry exhibits many of the cultural traits of these reform movements—holdovers, many of them, from the antebellum era before industrialization gained the upper hand. Like the Populists, Berry ranges widely in his complaints (though steering clear, to be sure, of nostrums like the free-coinage of silver). He frames issues in moral terms, often reducing them to the individual level. His hope is to appeal to citizens from all walks of life. And he dreams of a new order. His vision of success comes mostly from the past, and it is a vision distinctly familiar—the independent farmer or craftsman who plies his trade with little interference from industrial outsiders. This was the dream of many Populists, and agrarians today might dream of it still. The Populists rose and fell because their moral dreams lacked any means of accomplishment despite their mass organizing. Even as they worked in the political arena they were pulled back by their commitment to individual independence. With even less interest in politics, with less interest in targeted reform efforts, Berry seems even more detached from forms of power that might bring systemic change. Berry knows this, to be sure, and he has said as much.[18] Yet having said it, where are we left?

Individualism

Many of today's environmental goals—most of them, probably—are ones that require concerted action to achieve. The individual acting alone can accomplish little. Garrett Hardin highlighted one source of this predicament, which he termed the tragedy of the commons. Hardin's tragedy arose when users of a natural commons were able to increase their uses at will, exceeding the land's carrying capacity and causing degradation. What difference did it make if one person refrained from overusing the commons when someone else would do so? Matched with this ill is the tragedy of fragmentation, probably the greater tragedy in the United States today.[19] This tragedy arises when a landscape is fragmented into small pieces and no person or entity retains enough power to promote goals that require landscape-scale planning. Urban sprawl provides one example of this tragedy. The decline of connected wildlife habitats and of healthy rivers connected to floodplains is another.

If we want to point fingers, in terms of land-abuse causes, we can take note of the institution of private property, at law and (more significantly) in American culture. Secure private property can encourage a long-term caring attitude but it also accords individual landowners the power to degrade lands if they choose—neighbors be damned. The effect of private property as commonly understood is to vest so much power in individual owners that the community as such cannot coordinate activities at larger scales. Land abusers benefit from governments that govern least. Then there is the competitive market, which pressures participants to cut costs and, in the case of products from land, encourages or even compels landowners to work lands too hard. When the low-cost market leaders abuse their lands, pushing external harms on neighbors, future generations, and other life forms, other market participants can be compelled to follow suit or else lose out.

The point to emphasize is that individuals are subject to forces beyond their control. They live and work within systems that they did not make or choose. Their choices are constrained, particularly as laborers and producers. In addition, the economics of sound living are often skewed against moral behavior when the individual acts alone. Illustrative here is the individual who rides a bicycle to avoid polluting the air. The cost in inconvenience is borne by the individual alone (inconvenience and, in urban settings, breathing vehicle exhausts), while the air-quality benefits are enjoyed by everyone. The situation presents a mismatch in the allocation of costs and benefits. The game, thus, is heavily stacked against moral living. Were the rules of the game different, moral living could become far easier.

Like many critics, Berry implies that personal behavior is a matter of choice and that lifestyle choices reveal values; different personal values would yield different choices. The reasoning has merit, to be sure, and yet it shortchanges a critical point. We can see what is missing by returning to Hardin's tragedy of the commons, which he illustrated with the case of the hypothetical livestock grazer who continues adding animals to the commons even though overgrazing results. A grazer might well do this and at the same time support strict collective limits on overgrazing. The grazer who acts alone might degrade the commons; the same grazer, getting together with others to develop rules for using the commons, could readily vote to protect the land's health. The stances might appear inconsistent but they are not. In both cases, the grazer promotes his interest. The difference is that the grazer in one setting acts as isolated individual and in another as citizen or co-governor. As an individual the grazer can do little to protect the commons; as a member of a governing board she can do far more.

Berry's insights would be stronger if he took more note of writings on the tragedy of the commons. The farm landscape as he sees it is in important ways a commons and needs to be understood and managed as such. On the challenges of managing such a commons we have the useful work by the late political scientist Elinor Ostrom and others, explaining the necessary elements of successful commons management.[20] Sound commons management requires collective action through carefully crafted governance regimes; it cannot rely on individual morality. Basic writings on environmental economics provide further insights into the challenges of orchestrating good land uses at the landscape scale, writings on matters such as externalities, free riders, and discount rates. The problems are known, and the workable answers are all ones that require systematic changes put in place collectively.

Perhaps the central problem with Berry's stance, if we are to take it alone, is that it blames the individual as such for problems that are in fact far bigger than any individual, problems that permeate the institutional and cultural worlds in which individuals live. The failings of society, he seems to say, all lie at the level of individuals understood as independent moral beings. But individuals have no real choice except to live in the world as they find it, not just the natural world but the institutional and cultural world. Their lives, ways of thinking, and modes of daily existence are inevitably shaped by their times and contexts. Most people struggle to find and retain a satisfying, honorable place within the system. Few of them have the means or temperament to drop out and cut themselves off from the economic systems that foster degradation. Berry's portrayal of the individual poses a significant paradox. He tells us that all lives in a community are and ought to be highly interdependent. He honors the community as such. But when it comes to criticism, the interdependent system as such shoulders little blame.

What's missing here is a clearer recognition of real-world power, a clearer sense that most people get pushed around. Their options are greatly constrained in terms of employment and places and modes of living. An urban dweller needs electricity and typically has only one way to get it. A worker needs to get to a job site and a gas-guzzling car, a cheap one, may be the only option. Food buying for many people does offer possibilities for better living, and Berry mentions that with regularity: start by knowing your food sources and supporting local farmers.[21] But once that is done, what is next? Is local food buying a model that can be replicated in other aspects of life? Berry has often linked the local food economy in his home region with a locally based forest economy.[22] He convinces us that small steps can be taken in this setting by forest owners and small-scale loggers, at least when the lands include valuable hardwoods. But small-scale furniture making does not compete in the market with big-scale operations. Indeed, many US furniture makers have shifted operations to Asia. It is not reasonable to expect more than the tiny few to make a go of it in the handcrafted furniture market or to expect ordinary, cash-strapped consumers to invest in their durable wares. In the end, it is difficult to expand the example of the local food economy into other realms, or even to grow the local food movement to make a real dent in grocery-store sales. Solar-panel electricity generation comes to mind, and a few people might get by with woodstoves or other biofuel heat sources. But the further examples are not many.

Berry, in short, seems to overstate the power individuals as such have to make significant changes in their lives and ecological footprints. Putting such blame on them increases feelings of guilt and—human psychology being what it is—no doubt turns many away. To blame individuals when they in fact wield little power is to divert attention from larger failures of collective responsibility.[23] It is also to portray individuals in unduly harsh light, as being less morally responsible than they would be given better options.

Berry's stance, it should be clear, is strongly linked to the version of liberty that he embraces, the liberty that gained dominance in the time of Jackson, displacing the earlier, different liberty associated with civic republicanism (and much before it with Stoic philosophy). The new liberty stressed individual choice and freedom from interference, shorn of any duty to step forward and work with others for the common good.

As political scientist Kimberly Smith has said, Berry at bottom is a political libertarian, "deeply skeptical of mass politics and government institutions."[24] He resists the notion that government should tell people how to live, even as he laments that governments fail to protect small farm communities from the economic forces pushing them down. We hear no mention of, for instance, land-use regulation to control bad land uses. Or to turn to a particularly pertinent issue, Berry has little to say about water pollution caused by leaking household septic systems in regions such as his own. Faulty septic fields are a major cause of water pollution in rural Kentucky. The solutions, for once, lie within the powers of landowners. But libertarian culture resists the notion that government might inspect each home and demand that leaks be stopped. Where is the local political leadership that draws attention to such a problem and proposes binding solutions? Where are the cultural critics pointing out that political libertarianism, modern-day Jacksonian democracy, is in fact a big part of the problem?

Given Berry's recurring criticism of government, given his libertarian streak, we are left to wonder whether Berry does not in fact share significant traits with the actors whom he opposes, those who raise high the banners of economic liberty and private property in their too-successful efforts to ward off demands for accountability. At bottom, negative individual liberty is the liberty of the criminal class.

The Path Ahead

To see these limits on individual power and the many needs for collective action is to highlight where Wendell Berry's thought might usefully work in tandem with the ideas of others, with the conservation visions of observers who have thought more particularly about ecological economics, collective governance, private property, and the biology of conservation. Berry's strength is moral challenge, and he offers alluring visions of healthy communities of people living in place in ways that can endure. He sees clearly, and describes for all, the traits of the responsible landowner, bound to a place and devoted to its long-term health. What needs to supplement this is some version of civic republicanism, of ancient Stoicism, that accentuates collective self-rule and promotes structures of public decision-making that nurture the common good. Individual virtue is essential, and Berry is one of our most persuasive virtue ethicists.[25] But virtue needs to include a commitment to collective civic action. Berry's ending point, lacking any faith in movements and collective action, is that individual moral change will undercut the current system, as he explains in one of his late essays (quoting T. S. Eliot):

> If enough of us will accept "the wisdom of humility," giving due honor to the ever-renewing pattern, accepting each moment's "new and shocking/Valuation of all we have been," then the corporate mind as we now have it will be shaken, and it will cease to exist as its members dissent and withdraw from it.[26]

But how long must good people wait for this to happen? And, more pertinently, might organized efforts help bring this about by pushing not just for legal change but for better ways of thinking and talking?

The temptation, perhaps, is to say that we merely need to take Berry's moral prophesy and mix it with the kinds of practical organizing that today's environmental groups undertake. The combination, we might hope, could yield plentiful fruit. But we need to appreciate here what an awkward match this would be, particularly given trends in the conservation cause itself.

Decades ago the conservation movement exhibited a moral edge when it had bad actors to blame and it could attack them with vengeance—the polluters, poisoners, and irresponsible dam-builders. But those were the old days, when culprits were few and well known. The situation today is more complex. Particularly in the case of bad land uses (more so than climate-change gas emissions) the ills arise from widespread sources. The opposition is stronger, and for various reasons overtly moral language appears suspect. Dollars increasingly flow to land trusts that do little to ruffle feathers or cast aspersions. The way to improve land use, one now hears, is to buy land or pay bad-acting landowners to change their ways, not to criticize their bad actions. Not all of the environmental movement certainly, but much of it, has become tame, relying on vague, unobjectionable slogans ("connecting people to wildlife" or "conserving land for people"). The moral edge is worn down, goals are too modest, and cultural criticism is muted.

It is useful to lay this reality on the table because it explains why Wendell Berry, while widely read and admired, has had little direct influence on most environmental reform efforts except for those related to local and organic foods. Berry's strident moralism is well received by environmentalists personally but they are reluctant to use it themselves, not in public. They are reluctant to be openly critical except when targeting big businesses (the Keystone pipeline, for instance; the Blackwater Gulf oil spill). It is better, many sense, to work within current systems, to negotiate calmly for small advances—or as David Orr has put it critically, simply to walk north in a train rapidly heading south.

What this means is that Wendell Berry's clarion call for new ways of seeing land and people, his proclamation of land- and community-based virtues, proposes a significant shift in today's efforts to push for environmental reform. If we are to take Berry serious, as we should, his ideas need to be points of beginning for a new strategy to address misuses of nature and the culture and economy that stimulate and legitimate them. They cannot simply be tacked on to today's more constrained, technical environmental tactics. If good land use is possible only with an overhauled culture and with new heightened appreciation of communal ties and dependencies, then cultural reform needs to become element 1 of the long-term push.

The path to a place of greater hope is a path that entails changes in the ways we see and value nature and understand our place in its evolutionary web of interdependencies. On this point, Berry's message fits well with that of Aldo Leopold and with the lessons of ecology. As clearly as Leopold ever did, Berry has pointed his finger at ills in modern culture that simply must change if we are to live on land in ways that can endure, ways that promote Berry's one "absolute good."

We must begin by giving up any idea that we can bring about these healings without fundamental changes in the ways we think and live. We face a choice that is starkly simple: we must change or be changed. If we fail to change for the better, then we will be changed for the worse. We cannot blunder our way into health by the same sad and foolish hopes by which we have blundered into disease.[27]

Surely Berry is right in all of this, and his message of virtue and love needs to go out widely. But changing for the better might well mean shifting away from Berry's type of political libertarianism, mixing his love, as Reverend King urged, with organized power in pursuit of justice. The exemplars we now need are not just the honorable, responsible Epicureans that Berry so warmly presents but Stoic versions of them. We have blundered into disease with a worldview significantly shaped by economic liberalism. That part of our worldview might well need as much or more change as any other.

THREE

Impressionism and David Orr

On college and university campuses for nearly three decades, David Orr has been among the most visible, highly sought speakers addressing our somber environmental plight. He is a long-time educator himself—head of environmental studies and politics at Oberlin College—and knows how to engage and energize audiences. According to a recent biographical sketch he has served as scholar-in-residence so far at five universities and received numerous awards and prizes (including six honorary doctorates) for environmental leadership. He has also lectured at hundreds of colleges and universities throughout the United States and Europe and his audience halls are typically packed. Orr speaks with passion, his intellect ranges widely, and his brilliance is apparent. He urges readers earnestly to join him in working for a better world. For many listeners it is a message that brings fresh air and energy.

David Orr works hard as a writer as well as speaker. He favors the essay format and has penned dozens of them, full of sparkle and passion. Many have appeared in the half-dozen books he has authored over two decades. One of his publishers, Island Press, has released thirty-five of his best pieces under the title *Hope Is an Imperative: The Essential David Orr.* In preparation for this volume Orr revised many of his essays lightly, updating data, making illustrations current, and here and there expanding his arguments. The volume presents Orr's thinking in its considerable variety and richness. It is more than simply a handy source for accessing Orr's thought because Orr did the picking and organizing. The book is thus the most systematic presentation we are likely to have of his wide-ranging thought, organized as he would organize it.

Orr is worth considering here at length not just because of his individual visibility and the unusual breadth of his inquiry but because he draws in and exemplifies the third major strand of serious conservation thought. Leopold supplies the perspective grounded in ecology and (now) conservation biology; he is the advocate of wildlife, wilderness, and the functioning of ecosystems. Berry stands as a penetrating voice for pastoral landscapes, for communities of people living in place and working in concert to care for that place. He provides light for the local food movement, for the care of domesticated animals, and for the economic and cultural plight of those who strive to live on land in ways that sustain the land, themselves, and their families. To add Orr to this is to bring in more than a further voice of cultural criticism. Orr can stand for the strand of conservation thought that criticizes industrial methods and modes and that demands new and much better technology. He draws in the design, engineering, and planning professions—the design-with-nature methodology—and others that are sure human ingenuity, well aimed, can bring vast gain.

Giving Orr's essays a center is Orr's belief that, while our planetary plight is ominous and our prospects are dim, we nonetheless face an imperative to take action. We are charged to transform ourselves and the world around us, or collapse trying. Orr calls not for optimism—a mental frame, he asserts, consistent with inaction—but instead for hope, which in its authentic form, he relates, "comes with an imperative to act" (xix).[1] "Hope is a verb with its sleeve rolled up" (xix). "Hopeful people are actively engaged in defying the odds or changing the odds" (324). "Hope, authentic hope, can only be found in our capacity to discern the truth about our situation and ourselves and summon the fortitude to act accordingly" (332).

It is good that Orr has framed his essays this way because they could easily, read in series, promote despair. Some of the book's recurring bleakness has to do with flaws in modern culture; put sharply, as Orr does, we are a prideful, ill-behaving people, in need of a substantial overhaul in values, perceptions, intelligence, and spiritual core. Bleakness also hangs over our pressing problems, which are, Orr relates, deep-rooted in our values, economy, institutions, and landscapes. We can get atop them only by undertaking a radical transformation of human life as we know it through "deep, transformational change" (337), work that, once begun, could keep us laboring away for "several centuries" (90). Meanwhile, the opposition is well funded, potent, and too often ruthless; the "news" media are inattentive and gravely distended; and the forces of commercialized distraction can readily overwhelm. The odds, in short, are not attractive. Despite this darkness, though, Orr's volume is infused with a drive onward and upward and its overall effect, as he shares his passion and energy, is upbeat. In a brief foreword, Fritjof Capra describes the volume as "both wake-up call and inspiration." He is right, and right also in observing that, "what shines through [the essays] most of all is their author's deep passion for humanity and for the living Earth" (xiii).

There is much wisdom embedded in Orr's essay collection and Orr's work. As well as any of today's environmental voices Orr has been able to stand back and survey the entire scene. His messages are timely and sound: about the severity of our ecological ills and the lateness of the hour; about our need to think big; about the false allures of much that passes for reform; and more. He is rare among environmental writers in extending his reach so broadly and in embracing problems in their full social, intellectual, cultural, and technological complexity. He loves the world. He loves life. He wants to revive and enhance its health and beauty, for the good of humans and all other life. And he hopes his readers, like his live audiences, will share his passions.

Orr's writings are noteworthy for these various good traits. They are also noteworthy and invite study because they provide a valuable opportunity to study contemporary environmental rhetoric, identifying its strengths and limitations. Orr's essays have much to say about modern culture in terms of how it engages complex issues and struggles to address them. Orr knows he must speak and write in ways that draw attention, as he does: messages are unlikely to work unless they are heard. Yet, punchy, powerful messages have their known costs, key ones having to do with consistency, relative priorities, and systematic coherence. The environmental effort needs to come to terms with these costs, to understand them, and, insofar as it can, find ways to reduce them. Orr's writings, better than those of any other writer, provide the opportunity to do so.

Our environmental plight is extraordinarily complex, as Orr makes abundantly clear. Complexity, in turn, would seem to call for careful, systematic inquiry; it certainly does in scientific work. It calls for proceeding step by slow step, identifying the components of our predicament and crafting ways to remedy them. Perhaps our plight is best summed up as a single problem, as Leopold and Berry have described it: as an overall failure to develop ways of living on land without degrading it. Perhaps the better analogy is the ill patient who suffers from multiple disorders, each calling for attention. Even in the latter case, though, diagnoses of the multiple ills need to be assembled carefully and proposed treatments shaped to work together in combination. Whether suffering one ailment or many, the patient—the individual human, the culture, the earth—needs to be thought about as a whole, leading to an integrated strategy for change.

Taken one by one, as they were meant to be taken, Orr's writings work

well. Fit together, though, they illustrate—inevitably perhaps, given audience demands—aspects of modern culture that likely play roles themselves in fostering land declines and making reform more difficult. Among
them is the modern tendency to fragment big issues and to address parts of
them serially, in provocative ways that command attention. Orr is hardly
engaged in sound-bite dispensing, far from it. But his rhetoric is nonetheless contemporary in format and tone. Even the best current commentary
in leading journals of opinion draws readers using similar means of division and provocation. Our era shows little taste for systematic writers and
thinkers.

Fragmentation and severed connections are ills of the era. As Orr relates,
a long-term reform strategy is needed. To craft a sound one—a goal that
remains elusive—environmental reformers may need to confess that, both
in their advocacy practices and in the thinking that lies behind them, they
have not escaped these ills. The work of doing so continues on.

Where We Are

When portraying our dire plight on this thin-layered planet Orr speaks to audiences plainly and bluntly. We are not living well on Earth, he tells us, not
by a long shot, and we need to change our ways, fundamentally and soon.
Somehow, "fairly, durably, and quickly," we need to "remake the human
presence on Earth to fit the limits of the biosphere while preserving hard-
won gains in the arts, sciences, law, the open society, and governance, which
is to say civilization." This "is the challenge of our time, exempting no one,
no organization, no nation, and no generation from here on as far as one
can imagine" (xvi). Had we responded to problems when first told of them,
ordinary measures might have sufficed. But it is too late for that now. The
late hour calls for actions that are "extraordinary, unprecedented, and heroic
[and] at a scale sufficient to avert global catastrophe" (xv). And the challenges we confront go much deeper than simply technology choices. "They
are rooted in human nature, culture, history, and politics"; they are systemic,
and "can only be solved by changing the system" (55, 59).

As he presents his overview of today's landscape, Orr pulls no punches.
Clean coal, he taunts, is nothing but an "oxymoronic, people-killing, and
land-destroying absurdity" (3). Much that passes for conservative thought,
"disguised as a kind of super patriotism, . . . is merely reckless demagoguery
that serves corporate interests" (49). By and large people "are more ignorant than they are smart, and many seem to prefer it that way" (50). As
for stupidity, it is "randomly distributed up and down the socio-economic-

educational ladder" (52). Even tuition-paying students do not escape the lash. Too often they "resemble what Wendell Berry has called 'itinerant professional vandals,' persons devoid of any sense of place or stewardship, or inkling of why these are important" (256). And the conservation movement? It is "in failure mode" (60).

Orr presents his messages much as would a circuit-riding religious revivalist. He goes for emotional effect without slowing down to support his dire statements with the kind of detailed evidence needed to convince the doubtful. In his case the circuit-rider comparison is especially apt. Orr is the son of a Presbyterian minister and he uses biblical images and spiritual language to frame our plight. "Above all else," we are told, our downward slide "is a crisis of spirit and spiritual resources," and "something akin to spiritual renewal is the sine qua non of the transition to sustainability" (74, 71). Ecological design—Orr's all-encompassing vision of right living—"rests on the theological conviction that we are obligated, not merely constrained, to respect larger harmonies and patterns" in nature (186). "Can we," Orr muses, "build a sustainable society without seeking first the Kingdom of God, or some reasonable facsimile thereof" (88). "Can we harmonize the rhythms of this frail little craft of civilization with the pulse of the Great Heart of God?" (44–45).

Indispensable to Orr's evaluation of things is a factual and causal claim of central importance, one he wants audiences to hear exactly. Our environmental problems, Orr insists, are the results of human misconduct. They arise, pure and simple, because we humans are not living well in and with nature. Too often and easily we frame environmental ills in ways that implicitly portray nature as somehow deficient, as if nature was culpable for furnishing too little water in an arid land or a river could be faulted for excessive nitrates and silt. Orr wants none of this. We do not face water shortages; we suffer from wasteful practices and excess demand. Air is not polluted; people are polluting it. Volcanoes, to be sure, spew pollutants skyward, but such disturbances are natural occurrences; they are simply nature being nature. Our charge as humans is to find ways of living within natural systems that sustain their, and our, long-term health. Nature's modes of operating are realities that constrain our means of existence, even as we strive to make nature more hospitable. In sum, environmental problems are human actions that involve misuses of nature. To find their root causes we need to figure out who is doing the misusing and why they act as they do.

Orr's sure focus on human behavior supplies a much-needed corrective, particularly for environmental programs that place nature study front and center and that view natural scientists as appropriate program heads.

Environmental studies is not simply or chiefly the study of nature itself—what was once natural philosophy and then biology, zoology, botany, ecology, and microbiology. Environmental studies draws upon these scientific fields but goes well beyond them. At its center, the field places or ought to place the study of humans in nature: the study of how we affect nature and how nature in turn affects us. Also important the field includes (or should include) wide-ranging normative assessments of how we *ought* to be living, assessments informed by science but requiring an entirely different kind of intellectual inquiry. The scientific study of nature produces detailed depictions of nature and its functioning over time. Though indispensable, such science taken alone tells us nothing about whether a given landscape is or is not in "good" shape; it is says nothing about whether we are or are not living "right" in nature. Normative inquiries of this type—at the core, surely, of a sound environmental studies program—draw upon assessments of what is good and bad or wise and foolish, which is to say they are mostly not scientific.

Orr puts humans front and center in his essays because he wants to both figure out why we misuse nature and identify the changes required to live right. These are his central questions, just as they should be central to any program confronting our earthly predicament. Orr probes the structures and institutions of society and, finding fault in them, calls for fundamental change. But these structures and institutions are by and large reflections of who we are and what we know and want. Thus, to get to the true roots of our environmental ills, he suggests, we need to turn our gaze to human nature itself and to the cultures we have created. What is it about us, as limited beings, that leads us to cross the line between using nature and abusing it? Why do we so gravely misbehave, individually and collectively, in our natural homes? Without getting to the root causes it may not be possible to tame the economy and implement essential institutional reforms.

In essay after essay Orr considers these foundational questions of human behavior. His conclusions gradually accumulate and overlap, even as they display internal tension. We misbehave in nature, he explains, because we are limited, short-sighted, self-centered people guided by bad ideas and values. Too often we display an "inability or unwillingness to see what's right before our eyes" and are dragged down by "ignorance, garden-variety stupidity, and the tendency to put off to tomorrow what should have been done yesterday" (xvii). Put otherwise, "we are a flawed, cantankerous, willful, perhaps fallen, and certainly not entirely planet-broken race" (87). We are, in particular, weakened and threatened because "the powers of denial and wishful thinking cause us to underestimate the magnitude of our problems and overlook better possibilities" (68). Collectively we suffer from "a national epidemic of incoherence evident in our public discourse, street talk, movies, television, and music" (6). "Under the onslaught of commercialization and technology, we are losing the sense of wholeness and time that is essential to a decent civilization"; "we are losing the capacity for articulate intelligence about the things that matter most" (9, 6).

High among our unattractive features in Orr's portrayal is our "age-old addiction to force in human affairs," often manifest in "a drive to dominate nature that is evident in Western science and technology" (335, 81). We set ourselves up as lords and masters, not realizing that in the end we are obliged to live on Earth subject to rules we do not set. Collectively we are characterized also by ignorance and blindness, particularly to nature's ways and how we are inexorably degrading them. To live well we need ecological intelligence, "by which I mean a broad and intimate familiarity with how nature works" (167), an ability to become "mindful of the ecological fine print by which we live, move, and have our being" (205), a capacity "to comprehend patterns that connect, . . . getting beyond the boxes we call disciplines to see things in their ecological context" (169). "We've become," Orr laments, "a nation of specialists and technicians, not broadly educated and discerning people" (62). "The organization of knowledge by a minute division of labor further limits our capacity to comprehend whole systems effects" (16). Exacerbating this debilitation is our tendency to rush ahead, altering nature without fully considering the consequences. Our attention spans and time horizons are simply too short, our confidence in our cleverness much too great.

To understand our ecological ills we plainly need to learn more about how nature functions ecologically, particularly the basic process on which all life depends. But what kind of intelligence is needed to gain this knowledge or wisdom, and what role might new technology play in all of this? Orr is well known for guiding the design and construction of the ultra-green Adam Joseph Lewis Environmental Studies Center at Oberlin, a milestone in classroom and office-building technology. More recently he has brought the college together with the town of Oberlin to find ways to apply the underlying design and technology ideas in the local community, particularly to reduce carbon emissions. Orr regularly calls for new technologies that ensure the responsible handling of resources from cradle to grave. Yet, he is equally quick to point out that "more science and better technology won't be nearly enough" and he summons us to confront "technological fundamentalism" (xvii, 235). It is a dangerous myth, he tells us, to presume "that with enough knowledge and technology, we can, in the words of *Scientific*

American (1989), 'manage planet earth'" (239). Unless we become better people our ecological problems will "outrun our problem-solving capacity" (68). While technology is "sometimes a partial solution" it is also "a symptom of what ails us" and "decidedly not a panacea" (55). "Rather than technological breakthroughs," Orr states, "what we need, I think, is more like a homecoming that requires fewer highly paid experts and consultants, fewer conferences in exotic and expensive places, and more local knowledge, a more competent and empowered citizenry, and more reflection of what's important and what's not" (3). Put otherwise, "the barriers to a graceful transition to sustainability, whatever form it may take, are not so much technological as they are social, political, and psychological," firmly "rooted in human nature, culture, history, and politics" (67, 55).

As for our needed approach to learning, Orr has taken stances that both overlap and vary. On the one hand, he explains, we need "a larger and more rational rationality" and to cultivate "discerning intelligence in the public" (xvii, 61). Similarly, the shift to sound living will require "the integration of expertise across many disciplines" and "relentless analytical clarity" (210, 27). At the same time, though, Orr warns that our problems "are not solvable by rational means alone" (71). "We are at the end of our tether and no amount of conventional rationality or smartness is nearly rational enough or smart enough" (337). We require, along with rationality, a sense of wonder and awe in the face of nature's complexity. "Mankind will not perish," Orr observes (quoting Abraham Heschel), "for want of information; but only for want of appreciation. . . . What we lack is not a will to believe but a will to wonder" (42). Matched with this enlivened wonder must come a sense of gratitude, a commitment and capacity to address our problems not by "logic and method" but instead "by higher forces of wisdom, love, compassion, understanding, and empathy" (71). Meanwhile, we need to admit that "ignorance is not a solvable problem; it is rather an inescapable part of the human condition" (239). Thus, even as we seek to rise up "we cannot escape our creaturehood, and we can compound our problems many times over in the attempt to do so" (100).

Related to Orr's juxtaposition of sober, rationale inquiry with wonder and compassion is our challenge, as he sees it, of forging a temperament that helps us embrace our shortcomings and change our ways. Here, too, Orr's observations are insightful yet disparate. "We are in failure mode because . . . we are complacent and lack passion" (60), he sometimes explains. To be effective, protest is and perhaps must be grounded in anger. "Anger nourishes hope and fuels rebellion, it presumes a judgment, presumes how things ought to be and aren't, presumes a caring" (61). At the same time, we need to observe nature and be shaped by its teaching. We should mirror, in our demeanor and actions, "the things that one might imagine the Earth would teach us: silence, humility, holiness, connectedness, courtesy, beauty, celebration, giving, restoration, obligation, and wildness" (250). We should be mindful, too, that we are endangered by "a condition of spiritual emptiness" (68). We have stumbled because we have "failed to appreciate the depth of human needs for transcendence and belonging" (61). Our problems "can only be resolved by higher forces of wisdom, love, compassion, understanding, and empathy" which is to say "moral stamina" and a "higher level of spiritual awareness" (155, 71).

Can we, though, transform ourselves enough to become good citizens of the biotic communities that we help form? Can we rehabilitate ourselves? Orr's Presbyterian background leans toward a somber view of human nature and potential for change. Nonetheless he is guided by hope and, having traced our ills to human nature, is driven to assume that inner change is somehow possible. Thus, he informs us, a sustainable society "must be built on the most realistic view of the human condition possible" and "resilient enough to tolerate the stresses of human recalcitrance" (89). We must keep aware that "lurking in the shadows there is the darker side of human nature that can't be wished away" and remain mindful, as noted, that ignorance "is an inescapable part of the human condition" (xvii, 239). We are to keep aware, too, of the "corollary to the law of ignorance"—that "humans, in all ages and times, are inclined to be as unskeptical, gullible, and deceivable as those living in any other. Only the causes of our gullibility change" (51). At the same time, the more sanguine Orr confesses to "the quaint belief that if people only knew more, they would act better" (332). And he holds out hope for the reform effort that he terms "ecological design." Ecological design just might succeed in instructing us "in what we need and the terms of our existence on Earth"; it might "reorient our sense of time, giving greater weight to future prospects and to long-term ecological processes as well" (179) while vesting power in "people who know and value the positive things that bring them together and hold them together" (166). More than that, good design could "improve the integration of the human mind with its habitat and to fit in a larger order of things," enhancing "our collective intelligence by promoting mindfulness, transparency, and ecological competence" (177, 205).

Root Causes

Given Orr's emphatic and colored rhetoric it is perhaps unsurprising to find that his diagnosis of root causes shifts in focus from essay to essay. Most often, as noted, Orr spots the deepest roots in flawed human nature—in our knowledge, perceptions, yearnings to dominate, lack of passion (or of quiet), and the like. A particular claim sometimes made, already mentioned, is that our environmental plight is "above all else . . . a crisis of spirit and spiritual resources" (74). Personal and spiritual deficiencies, though, do not fully account for why we misbehave. Also at work are a constellation of forces related to weakened democracy, the decline of active citizenship, and the corrupting influence of corporate money. These are political problems, Orr explains. Because of them, "we do not have an environmental crisis so much as we have a political crisis" (62). "As much as some might wish it otherwise, environmental protection, climate stability, and the conservation of biological diversity are unavoidably political" (49). Powering these political crises is the "juggernaut of technology-based capitalism" (143) aided by "extreme interpretations of individualism and property rights [and] a pervasive suspicion of government" (114). "For nearly a half-century, government at all levels has been under constant attack by the extreme right wing with the clear intention of eroding our capacity to create collective solutions to national and global problems" (59). Such political problems call us to "repair and perhaps reinvent the institutions of democratic governance for a global world" (62).

Even as we admit these personal flaws and political ills, though, we cannot overlook the practical ways we interact with nature, individually and collectively—what Orr terms the issue of design. "Most, if not all, of our environmental problems result from poor design," he contends in the introduction to *Hope Is an Imperative.* "The 'ecological crisis' is the sum total of bad design with a tincture of bad intent, but the latter is not as easily solvable as the former" (163). The appropriate response, plainly, is "better design or what is coming to be known as ecological design" (xviii). Good design, though, does not come easily. A shift to it "will require a major change in how we think and so changes in education at all levels" (xix). Further— and to tie together politics and design—"ecological design also applies to the design of governments and public policies" (168).

Bad design is not unrelated to other root causes of abusive living. Science, understood as a way of interpreting and evaluating the world, also has something to do with it. "Modern science," Orr states, "has fundamentally misconceived the world by fragmenting reality, separating observer from observed, portraying the world as a mechanism, and dismissing nonobjective factors, all in the service of the domination of nature" (82). Similarly and gravely, we are dragged down by flaws in our economic modes of thought, academic and popular. "From the perspective of physics and ecology," we are told, "the flaws in mainstream economics are fundamental and numerous" (81). Economists err in "discounting future outcomes back to some purported net present value," a practice "geared to maximize short-term benefits, often at a substantial long-term cost" (115). Similarly, when "operating strictly within the boundaries of the neoclassical paradigm" economists "cannot account for the true costs of impaired security, health, beauty, and spiritual comfort" (47). Indeed, "the specialized language of economics does not begin to describe the state of our well-being, whatever it reveals about how much stuff we may or may not possess" (8). In some way we need to replace such reasoning with "better accounting tools that relate human wealth generation to some larger measure of biophysical health" (69). Further, we need to "change the timing of payoffs so that long-term costs are paid up front as part of the purchase price" (75).

Finally, there is the foundational issue that launched Orr's public career: the deficiencies in and of education, which exacerbate our declining awareness of nature and interdependencies and our weakened abilities to think and speak precisely. We have forgotten that "the purpose of a liberal education has to do with the development of the whole person," not simply imparting technical skills (273). Too many campuses are corrupted by a "marriage between the universities and power and commerce" (232). Ecological literacy is given short shrift nearly everywhere. Scant attention is paid to the physical campus itself as a teaching tool and blueprint of how humans should dwell in a place. At bottom, our education system is far from fostering, as it ought, "a sense of connectedness, implicatedness, and ecological citizenship" along with "the competence to act on such knowledge" (276).

The Backlash

Before taking up Orr's proposed solutions to our environmental ills, it is useful to take note of how Orr describes and explains the many elements of American society that do not see the world as he does. Orr is well aware that people who seek to promote healthy lands and communities have a fight on their hands. Day in and day out they "contend with competing political and economic interests," which have in recent decades "become rigid ideologies rooted in tattered beliefs that humans can do as they please with nature without consequences" (xvii). As he sees things, "the movement to preserve a habitable planet is caught in the cross fire between fundamentalists of the corporate-dominated global economy and those of atavistic religious and political movements" (58). Worse, since the election of 2008, "the forces of denial in the United States [have become] more militant and brazen than ever before" (58).

The environmental backlash—to use the phrase of historian Samuel Hays—has been around for decades. Orr is among many to rail against it and to accuse it of an array of character defects (greedy, short-sighted, religiously deranged). Americans generally take a head-in-the-sand attitude toward problems and needs for reform. The environmental backlash is more intentional and malignant than this. It seeks deliberately to confuse people about problems through misinformation campaigns while reaping profits from practices that are positively destructive.

One of the temptations, when writing about the backlash, is to present its stance simplistically and negatively. It is a temptation that Orr, among others, does not resist. To be sure, Orr is mostly preaching to the converted and might rightly assume that audiences will agree with his condemnations. But it hardly needs saying that people who see the world differently will not feel that their opposing views have been fairly presented, much less countered. The monied forces that promote confusion about human-caused climate change are very likely motivated in just the manner Orr describes. Against them, his labels stick. But what of the people who do not stand to cash in financially from climate-change confusion? And what of the intelligent people working for libertarian think-tanks and advocacy groups who honestly promote alternative perspectives? Can we dismiss them so glibly?

Orr's quick dismissal is partially due to his short-essay format and to the premium he places on passionate advocacy. Here the realities of audience and attention span come into play. For those seeking a fuller understanding of what is at stake, though, we need a more objective treatment. In truth, arguments can be made in favor of using popular economic techniques such as cost-benefit analysis to gain a relative sense of competing problems and such analyses require discounting of future costs and benefits. Many of Orr's calls for systemic change would necessarily require strong central governments, and critics might rightly fear that governments with the power to do what Orr wants would do a great deal of mischief as well. Many elements of neoclassical economics are soundly criticized (and have been, by Herman Daly and others), but there is truth also in the claim that free markets with their decentralized decision-making can handle massive amounts of information better than can organizational hierarchies. Not all natural resources, as they become scarce, can be replaced by others, but some can and have. New technology has significantly reduced many problems. The market does have built-in powers to stimulate activity that promotes improved environmental outcomes.

Orr might also note better that opposing forces in their rhetoric have ably deployed values and institutional arrangements that Americans hold dear, even as they also want clean water and abundant wildlife. When opposing forces draw upon these values and institutions they project public faces that are far friendlier than the evil money-grabbers that Orr sees and portrays. Americans value individual *liberty*, and many are convinced their liberties will suffer unduly should lawmakers undertake the ecological designs Orr has in mind. To many of them, the preservation of individual liberty might in fact be worth substantial ecological degradation. Similarly, well-designed communities and healthy landscapes can likely come about only through detailed, widespread land-use regulations, which seem to pose grave threats to individual *private property*. The environmental backlash in its property-rights form has powerfully argued that such land-use controls seriously threaten an institution that Americans hold dear. Is it a wonder that people listen and that the conservation movement suffers when it offers no response? The failings here, we should note, are not so much those of Orr as they are of the larger movement of which he is a part. Still, they stand as limits on the completeness of what Orr offers.

Much as it claims to protect liberty and property, the backlash effort has similarly latched on to the mantle of *democracy*, despite contentions by Orr and many others (based on sound evidence) that opposing powers are, to the contrary, succeeding in undercutting democracy. Here the backlash claim is that environmentalism, particularly nature preservation, is an elitist concern being pushed upon ordinary citizens by wealthy people who may or may not have to work for a living. According to one claim, environmentalism is a form of nature worship, a type of pantheistic religion. For government to embrace it, as people such as Orr seem to propose, would violate longstanding rules about the separation of church and state.

Orr does not take time to deal with these opposition claims. He does not acknowledge the rhetoric used by libertarians and Tea Party enthusiasts to resist the call for ever-more collective action to protect nature, people, and future generations. Can we, though, do what Orr wants—chart a path toward a greener future—without coming to terms with these backlash claims? Can we, for instance, realistically push for new land-use regulations without developing responses to widespread worries about violating property rights? Could we not usefully draw upon alternative strands of liberty, positive and collective strands, to respond more soberly to libertarian claims that individuals are free only when government leaves them alone? And could not environment defenders reclaim the mantle of *equality*, reframing it in terms of equal access to ecologically healthy homes and workplaces and to rejuvenating wild nature?

Arenas of Change

When it comes to reform proposals, Orr is not inclined to limit himself to ones that might succeed in coming decades. His hopes are bigger and stronger than that. His central reform thrust—to judge from the space he devotes to it—is for significant change at the individual level. Our ills are due to our intellectual, moral, and spiritual limits and defects. We thus need to improve and redeem ourselves, just as Wendell Berry has said. We need, in particular, to gain greater ecological literacy and become more attentive to connections and interdependencies. This includes a solid "comprehension of the interrelatedness of life grounded in the study of natural history, ecology, and thermodynamics" (258). We also need to regain a "discerning intelligence" (61), while broadening our vocabularies so we can think and talk clearly about the world and our place in it. As many people as possible should acquire a sound liberal education, the goal of which is "not mastery of subject matter but mastery of one's person" (243). Similarly, we would benefit from spiritual renewal if not rebirth to reconnect to nature and the sources of life.

Many of Orr's educational goals seem to require learning from scientific experts. The elements of thermodynamics, for instance, are not gained by chatting with friends at a coffee shop. At the same time, though, Orr embraces a strand of populist thought that honors knowledge acquired in other ways, by listening to neighbors or relatives and by direct engagement with nature. His varied perspectives here do not clash exactly, but they require some care to piece together. His proposal for education reform includes an expansion of the liberal arts curriculum to include "restoration ecology, agriculture, forestry, ecological engineering, landscape design, and solar technology" (250). At the same time he proposes that we get by with "fewer highly paid experts and consultants" (3)—a comment usefully placed alongside his Lewis Center project at Oberlin, which drew in the nation's best experts and consultants although without, perhaps, paying them highly. "We need to protect local culture in all of its forms" (11), he says, though presumably not including forms based on industrial agriculture, strip-mining mountains, and clear-cutting old forests. We need "to restore and preserve traditional knowledge of the land and its functions" (103), although again presumably not when it conflicts with the scientific lessons of wetland ecology and conservation biology. "All sustainability is local" (145), he asserts when talking about the benefits of protecting particular places but without contradicting his claims about needed national laws and global agreements.

A similar tension in Orr's thinking that is usefully recognized has to do with the pace or tempo of the reform effort. Orr makes clear that the hour is late and prompt action required. "As the years tick by, we are nearing (some say we have passed) irreversible and irrevocable changes in the oceans, atmosphere, soils, forests, and entire ecosystems" (xv). "The ideas that we need to build a sustainable society," accordingly, "need to be widely disseminated and put into practice quickly" (27). At the same time, he criticizes roundly the high speed of modern society and our tendency to act on information and technological options before thinking them through. Many if not most of our problems, Orr contends, "can be attributed directly or indirectly to knowledge acquired and applied before we had time to think it through carefully" (15). "The application of fast knowledge generates complicated problems much faster than we can identify them and respond," and "the only knowledge that we've ever been able to count on for consistently good effect over the long run is knowledge that has been acquired slowly through cultural maturation." To rely on slow knowledge, Orr says, "does not imply lethargy, but rather thoroughness and patience" (16). Thus, we need to deal with problems expeditiously and should, as Orr has done in Oberlin, insist on trying out the latest technological breakthroughs, yet at the same time keep aware of possibilities of error, questioning and testing as we go along.

Orr's educational reform ideas extend well beyond changes to the kind of liberal arts education taught at his home institution of Oberlin College. He wants to see ecological literacy introduced at every level of schooling, with ample opportunities for students to engage with local landscapes, to learn about local land uses, and to learn from naturalists, farmers, and others with hands-on experience. He presumably knows, though, that such changes will be a hard sell for local conservationists in the era of incessant standardized testing and given the overwhelming push to concentrate on basic skills. He likely knows full well also that college students today seek out schools where they have wide flexibility to pick and choose among their courses. Even elite schools compete for the best-credentialed students by touting their willingness to let students chart their own educational journeys. In that light, what chance is there to get colleges and universities to impose significant requirements for ecological literacy? And if the prospects are in fact as trivial as they appear, is it worth spending limited resources pushing for them?

Orr's reform plan gives a central place to the rapid embrace of new technology, notwithstanding his populist strand and his recognition that actions in haste are often misguided. Here again his Lewis Center project, with its state-of-the-art energy and waste-management systems, is emblematic. For sound reasons he exalts the ideals of cradle-to-grave and cradle-to-cradle resource use: resources organic in form and origin should be returned to their earthly graves to form new soil, while resources that are human-made and not biodegradable should be recycled for reuse as manufacturing feedstock. Energy providers should shift to renewable energy while buildings are redesigned to reduce energy loads. High-speed rail and other forms of mass transit should replace fuel-hungry, dirty automobiles. Throughout the economy and in daily life, new technology should reduce our planetary impacts, near and far.

Calls for green technology are, of course, commonplace these days. Few dispute them. The challenge comes at the next step, the step of identifying ways to get new technology implemented. Even the best technology accomplishes little when it is left on the shelf unused. In a few settings businesses themselves adopt new technology as a means of saving money or as a way of enhancing a favorable public image of responsibility. But far more businesses—whole industries—are resistant, as are most consumers in their daily lives. Without much better means of forcing adoption, new technology holds little promise. How then can adoption be compelled or induced or otherwise brought about?

On this question Orr fails to fill in the gaps that technology advocates typically leave. He recognizes that government must get involved: new laws or programs will be needed, along with effective enforcement. But how is this to come about, particularly given his comments about corporate dominance of government and the disengagement of citizens from active political involvement? Without a sound plan to achieve implementation, new technology is only a fragment of a solution, useless without the missing parts.

New technology can make our buildings perform better, but we will remain committed to high-energy, high-polluting lifestyles until we take the addition step of redesigning our communities and rearrange our land uses—at large scales, well above the individual city block—so that our transportation needs are much less. Cities and suburbs are mostly built around and for the automobile. Homes and work-sites are often far apart. Even simple needs can require traveling to stores miles away. Would it not be wonderful if we could walk or bike to places; if we lived near green belts; if we could gather in well-designed public spaces, mixed with nature, to socialize and enjoy simple pleasures? Would it not be wonderful if we could, collectively, undertake "smarter planning that creates proximity between housing, employment, shopping, culture, public spaces, recreation, and health care" (28)?

Visions of this type, of radically altered city plans and simplified life-styles, are alluring to many audiences, although by no means to all. No doubt ecological degradation would decline significantly if we could some-how find our way to this new mode of living. But we have here the same issue we have with the lure of new technology. How do we get from here to there? In the case of redesigned cities, the challenges are even greater since our streets, parking lots, buildings, and transportation systems are already in place and extraordinarily difficult to change except at very small scales. Set-ting aside the physical challenges, new-urbanist visions will require major new forms of land-use regulation. Inevitably they will (and do) collide with politics as usual in the land-use game. They will (and do) collide as well with the cherished institution of private property rights and with an ingrained popular disdain for government intervention. Even when government gets involved in the regulation business, it is, as Orr well knows, subject to cap-ture by the interests being regulated. What, then, are we to do?

This question is, in reality, pretty much a showstopper. And it is impor-tant to realize that Orr's Lewis Center does not provide a counterexample or evidence that he (and we) is on the right path. The Lewis Center was a free-standing building put up on land owned by a not-for-profit entity (Oberlin College). Orr and the other leaders wanted a green building, and they were able to keep government out of the way as they went about constructing it. (In the works are a few projects covering multiple parcels subject to mul-tiple owners and an effort [the Oberlin Project] to expand uses of carbon-reducing technology in the small community of Oberlin.) But we must be honest when thinking about the scale of individual redevelopment projects and also about the amount of charitable or public money being poured into them. These are pilot projects, valuable without question. But they only be-gin to address the question of large-scale implementation. Small projects by-and-large do not challenge the entrenched land-use system. They do not call into question the rights of private owners. They do not challenge the cultural assumption that land uses are typically best decided by owners themselves, subject only to public rules to limit spillover harms.

No doubt we are wise to push forward with new building projects. But before we view such projects as first steps along a path to new urbanism we need to see where that path heads next. We need some confidence that if we follow the path we will get somewhere good, and not end up with a few green buildings scattered in a landscape that remains auto-laced and degraded.

Similar questions about implementation, about political and economic realities, arise the moment we talk about achieving major changes in the economic activities that dominate our land uses—in agriculture, ranching, forestry, and mining. Wildlife populations almost everywhere are gravely disrupted, with countless species declining in numbers, due to habitat alteration. Intensive land uses also alter water flows that exacerbate flooding and droughts while degrading aquatic life with fertilizers, pesticides, mining leachate, and more. The entrenched interests here are powerful indeed. Forty years into the Clean Water Act era, for instance, most Midwestern farm states have still hardly made a start on forcing agriculture to reduce its polluted runoff despite plentiful knowledge on how it can be done. To call for change is easy; to imagine healthier, more attractive natural landscapes is easy. Getting there, given the realities, is the hard part. And it is important to note that the environmental problems here, like (in truth) most environmental problems, will not be meaningfully affected by actions that are or are not taken by individuals in their daily, personal lives. There is little the average individual can do in isolation to address these ecological ills, no matter how ecologically literate and spiritually aware. Indeed, perhaps we do a disservice to people to enhance their ecological awareness so that they become more aware of, and feel greater pain from, the destruction going on around them. Big industries respond only to pressure; pressure requires power; and power requires organization, leadership, and good strategy.

Orr's reform agenda includes changes in the ways extractive industries conduct their activities. He goes well beyond that, though, in proposing significant changes also in the way we think about corporations as legal entities and in the powers that they wield. He would have government revise statutes vesting corporations with power so that corporations are less able to engage in activities contrary to the public interest and public will. He wants corporations to exercise far less influence—perhaps none at all—in political campaigns, and to have less ability to dominate government and public policies in ways that frustrate the common good. This requires, Orr says (quoting Marjorie Kelly), that we "[challenge] "the divine right of capital," "enforce corporate charters," and "place democratic controls on corporations and the movement of capital" (62). It hardly needs saying that reform campaigns of this type are ambitious, as Orr knows. "In our own history," he relates, "progressive reforms far more modest than those necessary for sustainability have run aground on the shoals of corporate politics" (91).

It turns out that, when examined closely, Orr's reform agenda relies heavily on major changes in laws and public policies, which is to say action by government rather than by individuals and businesses acting alone. "The time has come," Orr tells us, "for a more thorough reconsideration of law, the rights of property, the public trust, and the human prospect" (136). "Only governments acting on a public mandate can license corporations and control their activities for the public benefit over the long term. Only governments can create the financial wherewithal to rebuild ecologically sound cities and dependable public transportation systems" (70). At the same time, though, Orr largely speaks ill of legislative efforts to date. "Environmental laws" in his view "seldom prevent or solve environmental problems. At best they render them somewhat more manageable while providing fertile ground for legal wrangling over the permissible rates by which the citizenry is poisoned and the land degraded" (133). As for "the hodgepodge of laws and regulations that govern chemical pollution," they "are easily corrupted and constitute no effective protection to human or ecosystem health" (113). (One cause of legal deficiencies, he suggests, might be "a legal community ignorant of the scale of environmental problems" [135].)

What we get, putting these statements together, is a vision not of more laws and regulations similar to those we now have but of laws that are stronger, clearer, and more effective in protecting the public interest. The vision—like many of Orr's visions—is plenty alluring. But how can we get there? What obstacles lie in the way, and what strategies might effectively confront them?

Orr's stress on legal and policy change, particularly at national and global levels, puts into a different light his other calls for personal change, spiritual renewal, and better education systems. If in the end the real work of environmental improvement can come about only through changes in laws and government, then how much effort should be put into these other goals? If people can be induced to support needed government action, do we care about their levels of intelligence, their verbal skills, and their state of spiritual renewal? And unless citizens do support new laws and policies—and the public officials willing and able to implement them—have we gotten anywhere if we improve our education systems and broaden public knowledge of thermodynamics?

Orr knows full well that American democracy is in trouble, both because of "the corrupting influence of money" (62) and because citizens are disillusioned and disengaged. "Political reform," he states again and again, "requires an engaged, informed, and thoughtful citizenry" (63). "The transition to sustainability will require a marked improvement and creativity in the arts of citizenship and governance" (69). Orr's vision here is of vast numbers of ecologically literate people coming together to change the system, rising up in some way to demand that business as usual come to an end. Again, the vision is appealing but the questions remain: How do we get there? Where do we start? Do we begin by studying spirituality books to help ourselves become more spiritually aware? Do we go to a local grade school and see if we can organize a garden for young students to work in as part of their education? Given the urgency of the situation, what is the best path to follow?

Questions such as these may seem petty or distracting, but the reality is that conservation-minded people have limited energy and resources. Powerful interests defend the status quo. The US Supreme Court is laden with Justices who, by most appearances, endorse policies that are profoundly anti-environmental. If we accept Orr's survey of root causes, we all are to blame for contributing to the current mess; we all need to rehabilitate ourselves. The visions of green living are appealing enough, but how can anyone escape feelings of despair? The job tasks listed by Orr are overwhelming. To be sure, it is comforting to know that someone like David Orr is on our side and doing all he can to raise public consciousness. But when we walk out of the lecture hall, variously depressed and fired up, which way do we turn?

Collective Action

This brings us to perhaps the most unsettling element of Orr's essays, his commentary on environmental groups and on their past and future roles in addressing our environmental ills. Green groups rarely appear in his writings, and then mostly (as in Wendell Berry's writings) to be criticized. One complaint is that they are guided by policy wonks: "too much of environmentalism is about data and numbers that have little resonance with the public" (52). A stronger complaint is that the movement (overall "in a failure mode") is detached and ineffective if not misguided. As a whole, he contends, the environmental effort "is fractured into different factions, groups, and arcane philosophies" and "has come to resemble the nineteenth-century European socialist movement that became bitterly divided into warring factions, each more eager to be right than right and effective." The socialists, when the time came for change, "delivered bolshevism, and the rest, as they say, is history." In comparison, Orr explains, "the Right generally suffers no such fracturing, in large part because their agenda is formed around less complicated aims having to do with power and pecuniary advantage" (60).

Though not without merit, these criticisms seem rather strongly phrased. Orr's comparison with socialism and bolshevism, if taken seriously (as perhaps they were not intended), is a blanket condemnation of the environmental effort. And as for factions within the movement, is it fair to note that

Orr himself is chiefly a solo actor, on the road giving his own version of what is right? His reform agenda is, without question, exceedingly complicated with nothing like the narrow focus of the political Right. Orr sees a need for a dominant green message, for "a new vision, a new story that links us to the planet in more life-centered ways," for "a positive strategy that fires the public imagination" (74, 63). All of this sounds fine, but the solution to a fragmented movement would seem to entail a call to unity, to agreement on the fundamentals, a call that one hears dimly if at all. Indeed, we hear very little about future needs for environmental groups and for citizens who support them.

Orr in his writings often cites authors from the past (Lewis Mumford is a favorite) but he makes little use of environmental history, particularly histories of the environmental movement. Orr knows well that we have made significant progress addressing some specific environmental challenges. Lead in the atmosphere is down sharply. Chemicals degrading the ozone layer are down as well. Common water pollutants from point sources have been significantly controlled. Air nearly everywhere in the nation is much cleaner in many (not all) ways. Sewage treatment operations are vastly better. Millions of acres are set aside as wilderness. National forests are much healthier, and wildlife refuges have a surer focus on protecting ecological integrity. If automobile efficiency standards are not as high as they should be, they are nonetheless much higher than decades ago, and catalytic converters have cut sulfur and nitrogen emissions significantly.

To cite this progress is to pose an obvious question. How did these improvements come about, and might successes of the past provide lessons for the future?

Orr is hardly alone—indeed, he perhaps stands alongside nearly all academics—in overlooking and implicitly belittling the work of citizen-fueled environmental groups as agents of change. Past progress would not have happened without the involvement of not-for-profit environmental groups. Indeed, it is difficult to find major reforms that took place without strong pressure from them. Their efforts, to be sure, have now run aground. They are countered today by an organized backlash movement vastly stronger than anything that existed when our major environmental laws were adopted. But green groups have been pretty much our only path to success in the past. Given that record, is it sensible to cast them aside and assign them no particular role in the work ahead?

Today's environmental groups are not nearly as effective as they could be and need to be. Orr is surely wrong in saying that they are dragged down by arcane philosophies of the type that divided nineteenth-century socialists.

(He is likely thinking about academic writing on environmental ethics, a body of literature with little influence on environmental groups.) A more apt complaint is that green groups do not take ideas seriously (as opposed to scientific facts and economic analyses). Publications by green groups rarely get beyond bumper-sticker rhetoric. Individually and collectively they have nothing in the way of clear overall vision of humans living right with nature. Orr is right when he says "the public . . . knows what we are against, but not what we are for" (63). The problem, though, is not so much arcane ideology as it is a failure to work together; it is the institutional reflection of America's commitment to liberal individualism, of flying solo.

The environmental movement is, as Orr says, badly fragmented. And it would surely do well to come together to present "a more coherent agenda" guided by a "coherent vision of common environmental good that is sufficiently compelling" (63, 114). But single visions and effective messages will not arise without a strong call for unity and without a willingness by thoughtful conservationists such as Orr to work with colleagues to find common intellectual ground. At the moment, we have a thousand points of light flickering in the dark. Orr's is one of them, brighter and better directed than most. But to succeed we need a beacon if not a highly focused laser.

Orr's agenda sometimes calls for coordination and focus, but says nothing about how to get there. "We" need to "take back public words such as conservative and patriot," he says (61). "We need a positive strategy that fires the public imagination" (63). But who is this "we"? At one level it is the American people generally; at another level, it is citizens especially concerned about ecological degradation. But the most important "we" is very likely a unified, well-considered citizen environmental movement. "We" are not going to get anything done without working together.

To work together means organization, it means leaders, it means think-tanks, it means strategy, it means above all recognizing and highlighting the vital role of an energized, unified environmental effort ready to wage battle in the harsh world of today.

Impressionism and Clarity

In the fall of 1946 Aldo Leopold was asked to prepare the conservation platform for a new political party. He responded with an agenda limited to two items, written on half a sheet of paper. If we did not put these two points front and center, Leopold said, everything else was to no avail.

Leopold's second point (as noted in chapter 1) was that conservation needed to promote the health of the land community as a whole, not flows of natural resources. Land health was the guiding normative light. His first point was that society needed to reconsider how the individual as such fit into the community in terms of rights and responsibilities, the individual landowner most of all. If society did not get these points straight, Leopold urged, nothing else really mattered.

David Orr is not inclined to organize or narrow his thought in the way Leopold did, even though he regularly uses "ecological design" as a catch-all term. Orr is better seen as standing in the important line that included Henry David Thoreau and Walt Whitman. He is master of the one-liner, he overflows with enthusiasm, and he is willing to tolerate (if not, as Thoreau and Whitman did, to exalt) his inconsistencies. Orr stands also in the line of utopian writers, despite his dark, prophetic side. The term he has chosen to capture his vision, as we have seen, is *ecological design*. The term has a tech-nological flavor and seems at first glance to refer to engineering, construc-tion, and perhaps land-use planning. But Orr very much "intend[s] the term design broadly" (163). "Ecological design is the careful meshing of human purposes with the larger patterns and flows of the natural world and the study of those patterns and flows to inform human purposes" (165). Our overall challenge—learning "how a species pleased to call itself Homo sapi-ens fits on a planet with a biosphere"—is, Orr says, "a design problem" (27).

Design includes everything having to do with our interactions with nature; it is "the effort to recalibrate how we build, grow, make, power, move, live, and earn our keep" (163). The goal here is not simply one of long-term prosperity for it includes a high aesthetic standard—"to work so artfully as to cause no ugliness, human or ecological, somewhere else or at some later time" (206). Further, the work of design is not only technical. "As a kind of storytelling, design is a celebration of the life that connects us with the nature of the places in which we live and work and grounds us in the still larger story of the human journey" (210). Design is a "healing profession," and "the result of good design is, in a word, health—both human and eco-logical" (207, 163). Even "loving children is a design problem" (172). Life in this world would be good indeed, including such elements as "better poetry," "neighborhood book clubs," "great pubs serving microbrews," and "no more wars for oil or anything else" (331). "At its best, ecological design is the ultimate manifestation of love—a gift of life, harmony, and beauty to our children" (179).

It is refreshing to have someone speak so vividly and passionately about a better world, particularly in times of intractable wars, political malfunc-tion, rising inequality, and economic malaise. Perhaps today's colleges are particularly in need of such visions as a counterpoint to the world of video games and online chatting. But utopian visions have their known downsides, and efforts to implement them have typically proved ill spent. One danger is that the reform agenda is so vast and labor sufficiently scarce that efforts are spread too thin to accomplish much. Another is that utopian advocates, once understood as such, are too easily dismissed as dreamers.

If Orr were categorized not as a writer but as a painter he would fit in well with the late nineteenth-century impressionists, who responded to the world around them by capturing the light, using colors boldly, ignoring sharp lines in favor of basic shapes, and infusing their works with feeling. Orr's paintings, one senses, would have colors applied not just with broad brushes but in thick layers with palette knives, reacting spontaneously to unfolding events as if painting *en plein air*. Working this way, he fits in well with so many other commentators on contemporary issues who similarly use strong rhetoric and bold strokes without taking time to gain sure focus. The result as art can be inspiring, and in Orr's case it is. But once inspired, audiences need visions with more clarity and fidelity. And where is that to be found? Practically speaking they also need a hierarchy of lessons; they need no more than a handful of top priorities that are then supported in carefully organized ways by subordinate points. When the main messages number in the dozens, it is easy to be confused by them. It was wisdom that Aldo Leopold knew and remembered, but he lived in a day when audiences were different and others knew the same lesson.

Orr is right and useful in looking to flaws in contemporary culture for the root causes of degradation. Others need to do the same, with the aim of moving beyond boldly colored reactions to images that are truer and clearer. On the other end, in terms of goals and how we distinguish use from abuse, here too greater clarity is needed. How should we define good land use at varied spatial scales so as to formulate usable tests and standards? With root causes clearly seen, and with standards of good land use on the horizon, the many remedial steps that writers such as Orr put forth can be better assessed in terms of their powers to address root causes and move us toward our normative goals. New ways of perceiving nature will no doubt be part of the remedial plan. So how should we talk about nature as a whole? How should we comprehend our place as living beings in the web of life? Education might be part of it, although that is less clear given that today's problems do not seem caused by any shortage of available knowledge; in any event, education can be designed specifically to counter the most important root causes.

In the end, effective messages will be clear and often repeated, and they will be pressed by citizen-supported collective action of a type that environmental writers today (following Berry's lead, some of them) too often discount or even belittle. It is not likely to work for each painter to work alone, recording personal reactions and to change their reactions with shifts in the light.

In too many ways the environmental reform effort in the United States, inside and outside the academy, is stuck in neutral. Environmental groups have yet to come together into an effective movement that responds to critics and presents the kind of coherent, compelling message that Orr deems essential. Too many key challenges, highlighted by Orr, are simply not being addressed. As for campus programs related to the environment, they have hardly started, most of them, to put into the picture real-life humans—the kind of cantankerous, misguided, spiritually hungry people that Orr seems to know so well. These are the people who need to be brought in to help, Orr tells us. It is all about citizen activism, and people working together to make a better world.

To reach this point is to get back to Orr's title, and to his claim that hope is a verb with its sleeves rolled up. Perhaps it is. It certainly needs to be. No amount of good science will carry us forward, nor can we expect political leaders to arrive, *deus ex machina*, to deliver us from fallen ways. It is up to us, Orr preaches. In that regard he is solidly right, and we would do well to listen.

FOUR

The Cosmos and Pope Francis

Faced as we are with global environmental deterioration, I wish to address every person living on this planet. . . . I would like to enter into dialogue with all people about our common home.

—*Laudato Si'*, ¶ 3

Many things have to change course, but it is we human beings above all who need to change. . . . A great cultural, spiritual and educational challenge stands before us, and it will demand that we set out on the long path of renewal.

—*Laudato Si'*, ¶ 202

Given the vast writings on environmental ills it is noteworthy how little of it focuses on what might to be termed the issues of beginnings and endings. Environmental problems arise when humans misuse nature. Conservation efforts are intended to halt that misuse, to align our dealings with nature on the sensible side of the use-abuse divide. The beginning issues are those that probe the root causes of land misuse in human behavior—why we act as we do. The ending issues bring together the many relevant normative factors, including those reflecting limits on what we know. Out of this latter multifactor, normative inquiry would come a sense of where sound land use ends and land abuse begins, tailored to particular settings and updated as circumstances and normative choices evolve.

Much of what is termed "environmental scholarship" arises out of efforts to understand nature and how we have altered it. This is the science part of the work, by and large. It yields factual information, which is then drawn upon through interpretive means to understand our planetary plight and to craft reforms. If we set this scientific, fact-generating scholarship to one side, looking only at the rest, at the writings that make use of it, one is struck by the overwhelming dominance of writing centered on possible solutions to our ills. The pattern makes sense at first glance: we want to reduce our ills so we speculate and argue on how best to do it. But the tools or means used to remedy environmental problems need to succeed; they need to be effective. And it only makes sense, as we compare alternative approaches, to assess them in terms of their likely effectiveness. Would they confront the causes of land abuse and would they steer our nature-use practices in better directions?

As for end points, it is typical to point quickly at sustainability or sustainable development or (more recently) ecosystem services as the desired goal. But one hardly needs to scratch the surface of these end points to expose their vagueness and often radical incompleteness. It is little wonder that the terms are interpreted in much different ways and that wide-ranging intellects such as David Orr and historian Donald Worster are prone to dismiss them.

As for the beginning issues, the writing here is in ways even less satisfactory. A once-common line of thought attributed land misuse simply to ignorance about the effects of various activities: if people only had the facts about what they were doing to nature they would change their ways. These days, that understanding is rarely taken seriously, so abundant is the counter evidence. With similar crudity basic economic models attribute bad practices to "imperfections" in the free market. Yet that story, while containing truth, is also far from adequate. Other references point to human greed or overconsumption or overpopulation. These points dig deeper, as do claims that our fundamental flaw is a failure to respect other animals or Creation as a whole. But even these points, sound so far as they go, do not quite grasp the complexity of the situation. Yes, we consume too much, yes, we ignore nature's limits, but why?

At some point, more observers need to admit that the topic is highly complex and lacks anything like a simple answer. Among the scholars taking broader looks have been our top environmental historians such as Worster and J. R. McNeill.[1] They offer more nuanced, multifactored accounts that place considerable emphasis on the cultural aspects—the issues probed by Leopold, Berry, Orr, and others—as well as on the material and technology aspects, especially the harnessing and use of fossil fuels. If we are to rely on these scholars, the best we have, we misuse nature for a mixture of overlapping and connected reasons, many of them cultural.

Cultural Ills and *Laudato Si'*

In the summer of 2015 the Bishop of Rome, Pope Francis, released his much-anticipated encyclical, *Laudato Si'* (more fully, *Laudato Si', mí Signore*, Praise be to You, my Lord), his major contribution to the worldwide discussion about nature and culture. The pope's letter, dealing with "the care of our common home," was typically described by reporters as an extended meditation on our environmental plight and its social justice aspects. As a few noted, however, the encyclical is more wide-ranging than that. It is a strong critique of modern culture overall, especially of market capitalism and its embedded assumptions and values. And it is a call for radical change in the trajectory of the contemporary world, even as it embraces science and many technological developments.

Before exploring the pope's encyclical it may help to distill the central cultural laments given prominence by Leopold, Berry, and Orr as a way, in effect, of setting the stage. In combination, their writings provide a perspective, an American one, from which to consider the pope's criticisms and to situate his calls for change. Their main points, as we have seen, are follows:

One recurring lament is that we humans have detached ourselves too much from nature in terms of our thinking and valuing. We act like conquerors of nature, to use Leopold's phrasing, rather than—as we should—plain citizens. To be sure, we are special forms of life. But we exaggerate our specialness. We exaggerate our ability to flourish in the long run while treating nature as simply so much raw stuff, available for our use and consumption.

As we look at nature, we are too prone to see nature in fragmented terms, to see it as a collection of pieces and parts, as raw materials or inputs for our activities. We overlook or discount the interconnections, interdependencies, and emergent properties of the natural world. The reality is to the contrary, Leopold asserted as the first lesson in his mature talks. The land is a community and we are embedded in it, as dependent on its healthy functioning as other life forms.

A third root cultural ill is that we are too confident in our knowledge produced by empirical data collection and analysis and too prone to push aside all wisdom not thereby supported. Particularly in public realms we abandon other ways of learning and experiencing. We overlook our ignorance, charging ahead as if our knowledge were sufficient if not complete.

Related to our mental and moral fragmentation of nature is our tendency to stress individual autonomy and liberation in ways that discount how people are, and need to be, bound to one another. We are too prone to look tolerably if not favorably on the vigorous competitive pursuit of self-interest, defined with little regard for the good of others, of nature, and of future generations. The problem is not simply the denial of community membership, of the type Berry and Orr have regularly decried. It includes the now-prominent tendency to shift normative choices to the individual level—to allow individual rural landowners, for instance, to define good land use as they see fit so long as they avoid immediate harm to others.

Leopold, Berry, Orr—they and others have commonly pointed a wagging finger at the contemporary tendency to dwell so much on the immediate and short-term with little concern for the future. Worries about future generations as such count for little, as do concerns about the long-term plight of a community as such. The chief land-use decisions are now made by market participants, by businesses, investors, and landowners (including farmers) who use their lands for economic gain. Making money is deemed legitimate if not obligatory, which is to say it is socially and morally acceptable to cut costs and otherwise compete in a market that dwells on the here-and-now. To admit that an individual market participant may have little choice but to play this economic game is not at all to absolve a society of its embrace of such a corrosive system.

Finally, there is, many say, a sense that modern culture is deficient in failing to reflect feelings of gratitude for nature's bounties and, in some versions, a recognition of the sacredness of the given world. Leopold's deep gratitude permeates nearly all of his writings, much as does his love of nature: he instructs by sharing his feelings as much as by telling. Orr on occasion draws on his religious heritage to speak in faith-based terms. Berry does so regularly. Opposed to a sense of gratitude is the sense of entitlement, dominance, and control without limits.

These cultural deficiencies are surely important as root causes of bad land uses; it is hardly possible, probing our overall patterns of land use, to explain them without these causal elements. These troubling cultural factors also play another, related role, for they similarly make it hard for people to recognize what needs to be done. By legitimating bad practices they make it hard to think clearly about our environmental plight and to change our ways.

To gain clear sight on how we ought to inhabit lands, as we already noted, we need to reflect seriously on the many normative factors that would inform an all-things-considered assessment. Out of that reflection needs to emerge at least a general line distinguishing land use from misuse. But to do that line-drawing, the labor involved needs to make sense to people. It must seem sensible that people collectively would formulate such an overall normative vision, one that could then shape policies and guide conduct. Further, it needs to seem reasonable, in order to produce such a goal, that people would come together in some manner to debate the various factors and options, which is to say sensible to think of themselves as an interconnected group that acts as such and pursues a shared goal.

In one way or another the basic cultural ills at issue here all link to an excessive and distorted view of the individual human, understood as an autonomous, preference-seeking individual who is free to pursue a self-chosen vision of the good life. When we hold up this autonomous individual human as exemplar, when we see society chiefly as a collection of such individuals, then it seems proper to permit each individual to make independent choices about such key issues as the moral status of other animals, the duties we might owe future generations, and definitions of land-use beauty. Too often and fully we place such issues within the personal sphere; they are matters of individual choice, matters on which individual opinions can and do differ. When that happens, it becomes challenging to find ways to talk about Leopold's land community and our membership in it. It becomes harder to get people to recognize the health of that community as a shared good. Indeed, so pervasive at times is the tendency to respect individual liberty and preferences that we struggle even to find a language that draws attention to what we share, beyond mere public safety and a functioning physical infrastructure.

Without a significant reshaping of this overarching individualism it is hard to see how we might materially reform our particular misuses of nature. It is hard to see also how we might similarly craft new policies and restructure institutions to foster and demand good land use by everyone.

The challenges here, to be sure, are profound. But what hope is there that individuals, embedded in systems that constrain and push, can independently embrace and embody new ecological values? If new values of the type Berry promotes require a person largely to drop out of the modern economy (giving up, for many of them, access to decent health care among other goods), what then has been accomplished? The family farm or forest, if there ever was one, will likely shift into the hands of an owner less limited by moral qualms. Is this progress?

Enter the Pope

The papal encyclical, *Laudato Si'*, is a rare if not unique pronouncement, coming as it does from a major public figure. One is hard-pressed to cite a comparable document ever issued by a prominent public figure or, indeed, by any organization. Other officials and entities have taken on our global environmental plight. But even drawing together such international documents as Our Common Future (World Commission on Environment and Development, 1987), Agenda 21 (Earth Summit, 1992), and the Millennium Ecosystem Assessment (a report to the United Nations, 2003–2005), the resulting compilation falls short of the pope's encyclical when assessed as a survey of the root causes of our ills and an identification of the changes now required. For precedents written in English it is noteworthy, as we shall see, that the pope's central claims align reasonably well with the main messages of Aldo Leopold on the land as a community (Leopold's term), on the normative primacy of community health, on the need for collective action for change, and on the root causes in modern culture. Much of the pope's critique overlaps also with the wide-ranging claims by David Orr, considering his writings also as a whole, although Orr (like Leopold) has said far less about social injustice and the poor and lays less stress than the pope on collective responses. Most extensive is the overlap with the many volumes by Wendell Berry—hardly surprising, given Berry's own use of religious language and the Christian core of his understandings. Again the overlap is far from complete: Berry's complaints use local examples, not global ones, and he rarely addresses poverty directly. He differs more substantially by his dissent from calls for collective action. But the similarities nonetheless are striking, making the differences all the more noteworthy.

Laudato Si' is worthy of close study and great respect, even apart from the honor due the office of its author (who was aided no doubt by others). Taken as a whole, it is penetrating commentary on the trajectory of modern society, attributing both our misuses of nature and global poverty to deep-rooted strands of modern culture. In the pope's view, change needs to occur at that level; to act otherwise is to resort to superficial tinkering that may alleviate symptoms temporarily but leave the underlying disease to worsen.

True intellectual respect, of course, includes critical engagement. The pope fails to develop several key points as fully as one might like: his proposed reform of private property, for instance, and how he would likely synthesize his many comments on moral value and ethics. We need to hear more on them. The pope leaves unacknowledged the dangers and mixed record over time of encouraging readers to discern for themselves God's ultimate purpose, for us and the planet. Here we need to raise a red flag. On the issue of population, he rightly challenges those who push the concern without at the same time questioning their own ways of living. Yet his counterthrust, effective though it is, hardly responds to the merits of concerns about overpopulation and the church's role in sanctioning it. Finally, the pope fails to clarify a key question that permeates his work: whether further environmental abuse might be justified in the name of helping the world's many poor. Still, his contribution is as substantial as it is provocative.

The World and How We Know It

Rather than a problem to be solved, the world is a joyful mystery to be contemplated with gladness and praise.

—*Laudato Si'*, ¶ 12

Pope Francis addresses his encyclical to all people of the world, not to Catholics alone. As he repeatedly stresses in it, his aim is to put forth an all-things-considered assessment of our planetary plight, drawing on science, philosophy, politics, and economics, as well as Christian teachings. His is not simply a religious view, much less a narrowly Christian one. He hopes his encyclical will be accepted in that ecumenical way, although likely realizing that many readers will dismiss his comments as merely religious or Catholic, or, in the case of academics, sufficiently tainted by religion that even the overtly nonreligious parts of the document (the vast bulk of it) can be safely ignored. "I urgently appeal," Pope Francis pleads, "for a new dialogue about how we are shaping the future of our planet. We need a conversation which includes everyone, since the environmental challenge we are undergoing, and its human roots, concern and affect us all" (¶ 14). The religious perspective, he insists, is usefully added to other ways of knowing nature and obtaining guidance on how we live.

To aid readers Pope Francis begins his encyclical by introducing his dominant themes, ones that will recur like musical motifs as his eclectic composition unfolds. He will return repeatedly, he explains to readers (¶ 16), to

- The intimate relationship between the poor and the fragility of the planet;
- The conviction that everything in the world is connected;
- The critique of new paradigms and forms of power derived from technology;
- The call to seek other ways of understanding the economy and progress;
- The value proper to each creature;
- The human meaning of ecology;
- The need for forthright and honest debate;
- The serious responsibility of international and local policy; and
- The throwaway culture and the proposal for a new lifestyle.

Significantly, not one of his themes is phrased in specifically religious terms, and his stances on them are all ones that secular writers have to varying degrees also endorsed.

Interconnection. The pope's second listed theme, on the interconnection of everything in the world, introduces his holistic view of the world. At times (as in the first item on his list) he refers to the planet as such; at other times he refers to "the universe as a whole, in all its manifold relationships" (¶ 86); sometimes, more warmly and personally, he speaks simply of "our common home" (¶ 1). It is noteworthy that, in presenting this perspective, he avoids using any holistic metaphors that might draw objection, metaphors of the type that Aldo Leopold freely deployed as a way to connect with his audiences. He does not speak of the planet, or indeed of any landscape or ecosystem, in organic terms, likening it as many have to an individual organism. Similarly, he does not use mechanical metaphors such as those Leopold used when speaking with farmers accustomed to tinkering with tractors. The planet is not literally an organism, nor is it a machine. The pope's careful word usage here implies familiarity with the contentious debates on these metaphors and a desire to avoid getting ensnarled in them.

Instead of such holistic metaphors Francis repeatedly uses the language of interconnection and interdependence. He speaks of relationships among nature's parts and of the functional systems that in combination are termed ecosystems. "We need to grasp the variety of things in their multiple relationships," he tells us; to see "the interdependence of creatures" and to recognize that "no creature is self-sufficient"; to realize that "creatures exist only in dependence on each other, to complete each other, in service of each other" (¶ 86, citing prior Church teaching). Throughout the world "we can find any number of constant and secretly interwoven relationships" (¶ 240). "The laws found in the Bible dwell on relationships, not only among individuals but also with other living beings" (¶ 68). The universe is "shaped by open and intercommunicating systems" and we can discern by study "countless forms of relationship and participation" (¶ 79). Our knowledge of these links and lines of dependence could and should be better, the pope says. "Ongoing research" could "give us a better understanding of how different creatures relate to one another in making up the larger units which today we term 'ecosystems'" (¶ 140).

This interconnected natural whole is not simply the place where we live. We are embedded in it, a part of it, and counted among its vast array of living creatures.

> When we speak of the "environment," what we really mean is a relationship
> existing between nature and the society which lives in it. Nature cannot be
> regarded as something separate from ourselves or as a mere setting in which
> we live. We are part of nature, included in it and thus in constant interaction
> with it. (¶ 139)

In Aldo Leopold's terms, as we have seen, we are plain members and citizens
of the land community, not conquerors of it.

Widespread value. This planet, this universe, this common home is infused
throughout with value, Francis tells us again and again, using value to mean
intrinsic value, not just instrumental value to humans. Individual creatures
as such, all of them, have some form of value. "We are called to recognize,"
he states, "that other living beings have a value of their own. . . . Each creature possesses its own particular goodness and perfection" (¶ 69, citing prior
Church teaching), "its own value and significance" (¶ 76). "Each creature
has its own purpose" and "none is superfluous" (¶ 84). Individual creatures
possess this value indirectly as well as directly, by virtue of participating in
ecosystems that also have "intrinsic value independent of their usefulness"
(¶ 140). "Because all creatures are connected, each must be cherished with
love and respect, for all of us as living creatures are dependent on one another" (¶ 42). "Each organism, as a creature of God, is good and admirable
in itself; the same is true of the harmonious ensemble of organisms existing
in a defined space and functioning as a system" (¶ 140).

The pope's recognition of widely spread value also extends to species as
such and to future generations of humans. "It is not enough," he instructs,
"to think of different species merely as potential 'resources' to be exploited,
while overlooking the fact that they have value in themselves" (¶ 33). Similarly, "the notion of the common good . . . extends to future generations"
(¶ 159). "We can no longer speak of sustainable development apart from
intergenerational solidarity. Once we start to think about the kind of world
we are leaving to future generations, we look at things differently; we realize
that the world is a gift which we have freely received and must share with
others. . . . Intergenerational solidarity is not optional, but rather a basic
question of justice" (¶ 159). For Pope Francis as for environmental philosophers such as Bryan Norton, a focus on future generations can reorient how
we think of ourselves and why we live:

> When we ask ourselves what kind of world we want to leave behind, we think
> in the first place of its general direction, its meaning and its values. Unless we
> struggle with these deeper issues, I do not believe that our concern for ecology

will produce significant results. But if these issues are courageously faced, we are led inexorably to ask other pointed questions: What is the purpose of our life in this world? Why are we here? What is the goal of our work and all our efforts? What need does the earth have of us? (¶ 160)

These various claims made by Francis about the existence of value in the world are all ones commonly encountered in environmental writing. As such, they are accepted as legitimate views, even if open to contest. The pope, as expected, links this value to the presence and will of God, adding a religious backing to the stances. There is little reason, though, to challenge his stances on that basis—that is, for the nonreligious to reject them—given that the secular writers who similarly recognize moral value ultimately have little in the way of backing for their views other than sentiment and ecological facts, augmented and fleshed out by logic. The pope's stances could easily be put forth alternatively as "self-evident" truths, to use the language of Jefferson.

How we learn. The pope's view of the world shows its religious side also on questions of epistemology, on how we go about gaining knowledge or insight into the world. He is more prone than many others—than scientists acting as such, in particular—to remain open to all possible sources of knowledge and to show respect for the inherited wisdom of the ages, even if it has not been confirmed by empirical inquiry. "Given the complexity of the ecological crisis and its multiple causes," Francis tells us,

> we need to realize that the solutions will not emerge from just one way of interpreting and transforming reality. Respect must also be shown for the various cultural riches of different peoples, their art and poetry, their interior life and spirituality. It we are truly concerned to develop an ecology capable of remedying the damage we have done, no branch of the sciences and no form of wisdom can be left out, and that includes religion and the language particular to it. (¶ 63)

"We urgently need a humanism capable of bringing together the different fields of knowledge, including economics, in the service of a more integral and integrating vision" (¶ 141). "It cannot be maintained," the pope states toward the end of his encyclical,

> that empirical science provides a complete explanation of life, the interplay of all creatures and the whole of reality. This would be to breach the limits imposed by its own methodology. If we reason only within the confines of

the latter, little room would be left for aesthetic sensibility, poetry, or even reason's ability to grasp the ultimate meaning and purpose of things. (¶ 199)

These papal statements on other sources of understanding, it needs emphasizing, do not pertain (or so it seems) to ways of gaining factual information about nature and its functioning. On such matters, the pope supports the use of scientific methods, even as he admonishes scientists collectively to show greater humility. His emphasis on other ways of knowing relates instead to questions of value, ethics, purpose, and calling. Knowing blends into feeling, particularly feelings of love and gratitude of the types strongly urged by Leopold, Berry, and Orr. On this issue the pope turns to his predecessor, Saint Francis of Assisi, "whose name I took as my guide and inspiration" and "the patron saint of all who study and work in the area of ecology" (¶ 10). Francis of Assisi stressed that "integral ecology calls for openness to categories which transcend the language of mathematics and biology. . . . His response to the world around him was so much more than intellectual appreciation or economic calculus, for to him each and every creature was a sister united to him by bonds of affection." Without such affectionate bonds to nature, the pope urges, without senses of value that arise through an "openness to awe and wonder," we cannot situate ourselves rightly in our planetary home. "If we no longer speak the language of fraternity and beauty in our relationship with the world, our attitude will be that of masters, consumers, ruthless exploiters, unable to set limits on their immediate needs." This perspective simply "cannot be written off as naïve romanticism, for it affects the choices which determine our behaviour" (¶11).

In the view of Pope Francis, these essential emotional bonds to nature—bonds that Ralph Waldo Emerson similarly raised high in his famous essay "Nature"—become stronger and more persuasive when they are also grounded in religious faith. One way to appreciate the "infinite beauty and goodness" of nature is, as Saint Francis urged, to see it "as a magnificent book in which God speaks to us and grants us a glimpse of his infinite beauty and goodness" (¶ 12). "We are free to apply our intelligence towards things evolving positively" (¶ 79), which is to say to use and rely upon science to gather and monitor the facts. Where faith comes in is at the next step: "Faith allows us to interpret the meaning and mysterious beauty of what is unfolding" (¶ 79). In words that could almost be drawn verbatim from Emerson, Pope Francis urges that the "contemplation of creation allows us to discover in each thing a teaching which God wishes to hand on to us" (¶ 85). By "paying attention to this manifestation," by absorbing the teachings embedded in nature, "we learn to see ourselves in relation to all creatures" (¶ 85).

A vision of right living. Employing these many ways of knowing nature and connecting to it emotionally, the pope puts forth, serially in many parts of this encyclical, his overall normative vision on how we ought to live in this finely crafted world—this planet, this universe, our home. He does not synthesize his observations in the manner of Leopold and Berry; he does not propose a guiding vision such as "land health" (Leopold) or "the absolute good" of the "healthy community" (Berry). But his many comments can be brought together, even as he leaves open (as do Leopold and Berry, in their ways) how the goal of ecologically healthy landscapes fits together with calls to respect other species and individual creatures as such—with, as Leopold put it, the "biotic right" of other life forms to endure.

Linking all of the pope's normative comments is the overarching ideal of caring for the planet as such. As Naomi Klein pointed out in an early comment on *Laudato Si'*, the pope makes almost no use of the word "stewardship," a term implying duties as such. He relies instead and repeatedly on the notion of care, a more inner-guided emotional bond based on love and respect. Even so, his encyclical is not short on senses of duty and responsibility. Many of his statements relate broadly to our role in caring for the planet as such, treating it as a gift entrusted to us that we tend and pass along to later generations. "The natural environment is a collective good, the patrimony of all humanity and the responsibility of everyone" (¶ 95). "By virtue of our unique dignity and our gift of intelligence, we are called to respect creation and its inherent laws" (¶ 69). It is simply wrong, Francis states, to think that we enjoy absolute dominion of other creatures by virtue of "being created in God's image and given dominion over the earth" (¶ 67). To the contrary, "the biblical texts . . . read in their context" tell us that we are to till and keep the garden of the world, and keeping "means caring, protecting, overseeing and preserving" (¶ 67).

Pope Francis, however, goes beyond these general statements of caring to present a more detailed ecological vision. We are admonished to avoid causing the loss of any species (¶ 33). We are also called to show respect for the relationships among living creatures (¶ 68) given that "everything is interconnected" (¶ 70). "Renewal entails recovering and respecting the rhythms inscribed in nature by the hand of the Creator" (¶ 71). When more specific, Pope Francis, like Leopold, turns to the language of ecological function and systemic resilience. We depend on nature's ecological systems—its services as they are often termed, although not by Francis—in order to live and thrive.

Although we are often not aware of it, we depend on these larger systems for our own existence. We need only recall how ecosystems interact in dispersing carbon dioxide, purifying water, controlling illnesses and epidemics, forming soil, breaking down waste, and in many other ways which we overlook or simply do not know about. . . . So, when we speak of "sustainable use," consideration must always be given to each ecosystem's regenerative ability in its different areas and aspects. (¶ 140)

We need to live, then, so as to sustain the ability of nature to continue functioning in these ways, beneficial to us and to other life forms—a distinctly functionalist normative stance, much like Leopold's land health. At the same time, as noted, we must avoid causing the extinction of other life forms and recognize and honor the value in all other creatures.

This normative vision of Pope Francis expressly (and repeatedly) links the good care of nature with justice among people. The dual challenges are intertwined, he says, and cannot be addressed in isolation. His overall normative vision thus gains further clarity when he adds that each individual human enjoys a right of access to safe, drinkable water (¶ 30) and that each human is "endowed with unique dignity" and harbors "a right to life and happiness" (¶ 43). Francis adds detail to this indirectly when he identifies the major environmental ills that the world faces today. They include, of course, issues of pollution, toxics, waste disposal, climate change, the degradation of marine life, excessive use of monocultures, wetland loss, and similar concerns (¶¶ 20–21, 23–26, 39–41). Beyond that, the list includes the "disproportionate and unruly growth of many cities, which have become unhealthy to live in, not only because of pollution caused by toxic emissions but also as a result of urban chaos, poor transportation, and visual pollution and noise" (¶ 44). His normative vision, that is, includes sensible land-use patterns. It also includes access to nature by all, which means "sufficient green space" in cities to allow "physical contact with nature" (¶ 44) and "access to places of particular beauty" (¶ 45), along with protection of "those common areas, visual landmarks and urban landscapes which increase our sense of belonging, our rootedness, of 'feeling at home' within a city that includes us and brings us together" (¶ 151).

With this overall, holistic, or communitarian vision of flourishing life on the planet Pope Francis shares a vision that lines up with Catholicism's longstanding links to the natural-law tradition, even though Francis makes no mention of it. The healthy flourishing of the entire community of life is, for Francis, the inner or inherent ultimate good built-in to nature, the kind of ripeness to which the community would naturally trend if not thwarted or pushed off course. Humans, he says, are called to help promote this long-term development. At the moment we are not heeding that call.

Root Causes Revisited

The problem is that we still lack the culture needed to confront this crisis.

—*Laudato Si'*, ¶ 53

The most remarkable aspects of *Laudato Si'* emerge out of the pope's harsh indictment of modern culture, his insistence that humans alone are to blame for all environmental ills, and his omission of any possible excuse or alternative hope that might mitigate our culpability. He agrees that we have brought on our environmental problems ourselves, and all are accountable for them.

· By implication we cannot diminish our culpability by claiming that we are fallen or otherwise fallible creatures, unable given our inherent limitations to do better than we have (a line of thought prominent in much Protestant Christianity).

· We cannot plead that our patterns of using nature, however troubling to us, might in fact comply with God's inscrutable plans (again, a line of religious reasoning not typically associated with Catholicism).

· We cannot—again by implication—claim that we might rightly focus our energies on another, heavenly realm or on preparation for an afterlife, as if our time on Earth was unimportant.

· Finally, we cannot hold out hope for divine intervention of any sort, whether a Second Coming or Rapture or some divinely wrought termination of the world as we know it.

Our backs are against the wall, and we are accused of serious wrongdoing.

At bottom, the pope's attribution of responsibility to humankind—his strong criticism of our misdeeds—nonetheless reflects an underlying optimism and confidence in our species. We have the ability by using our God-given reasoning powers, he tells us, to figure things out. We can come to know what is right and ultimately do it. "Human beings, while capable of the worst, are also capable of rising above themselves, choosing again what is good, and making a new start, despite their mental and social conditioning" (¶ 205). Goodness may be long in coming, perhaps far beyond any time frames we ever employ, and further vast, needless suffering lies ahead.

But in the fullness of time it will happen. On this point also, on the pope's ultimate confidence in humanity, we see a line of thought more Catholic than Protestant. "No system can completely suppress our openness to what is good, true and beautiful, or our God-given ability to respond to his grace at work deep in our hearts" (¶ 205). No system, not even the arrogant, destructive one that we now embrace.

As for the root causes of our land abuse, Pope Francis lines himself up with those (like Leopold, Berry, and Orr) who trace our behavior to cultural flaws even as his phrasing and emphases differ. He is not unmindful that we use new technologies to degrade nature. But he is unwilling to let us blame the technologies themselves. We created them, he implies, and can and should control them. He does offer a few words about human population and its causal role, but as noted below he casts the issue in a different light and gives the concern (too) little attention.

The cultural ills that Francis sees might for convenience be organized into four categories, to which we can append the various flaws in our governing institutional arrangements that incorporate and entrench these cultural ills. They are usefully set forth rather fully, even as they overlap with points already made by others. For the pope, our basic flaws have to do with (1) our ways of perceiving and valuing nature and comprehending our place in it; (2) our sense of ourselves as autonomous individuals who can legitimately give primacy if not exclusive attention to our personal preferences; elevating them, when it comes to nature, over virtues and the common good; (3) our ways of knowing and thinking based on empiricism, instrumental, means-ends rationality, and the fragmentation and specialization of knowledge; and (4) our guiding myths of power, progress, and salvation by technology. These cultural flaws in turn have fostered and defended arrangements of power—economic, political, and social—that distort understanding and discussion and that thwart well-meaning efforts at reform.

Our place. Pope Francis does not turn away from the longstanding views of Catholics (and Christians generally) on the specialness or exceptionalism of human life. That understanding largely remains in place. Our form of dignity is "unique" (¶ 43), the pope tells us. Still, this specialness does not mean that only humans have moral value. As already noted, all living creatures have value as such, and Francis in various places includes humans within the category of living creatures. He emphasizes further that, like other creatures, we are embedded in ecological systems that have value, and that our links with other life forms have moral implications to them. Humans remain special, but they are less special than they commonly think.

The pope is particularly pointed in challenging widespread beliefs that place humans too far above or apart from other species. "Modernity has been marked by an excessive anthropocentrism," he laments (¶ 116). "Clearly, the Bible has no place for a tyrannical anthropocentrism unconcerned for other creatures" (¶ 68). Similarly, Catholic teaching "clearly and forcefully criticizes a distorted anthropocentrism" (¶ 69) and the "ultimate destiny" of the universe, in the fullness of God's time, further argues "for rejecting every tyrannical and irresponsible domination of human beings over other creatures" (¶ 83). A failure to honor the value of other living things, and the relationships among them, does more than disrespect other creatures. "Once the human being declares independence from reality and behaves with absolute dominion, the very foundations of our life begin to crumble" (¶ 117). "A misguided anthropocentrism leads to a misguided lifestyle" (¶ 122). Put in more religious terms, our error comes from "presuming to take the place of God and refusing to acknowledge our creaturely limitations" (¶ 66).

Limits on autonomy, social and moral. Our deficiencies in thinking about our place in nature—our limited exceptionalism—are linked to defects in the ways we think of ourselves, chiefly as autonomous individuals who can charge ahead, pursuing self-centered interests, with little regard for the common good and future generations. Several of the pope's comments on this topic repeat charges commonly heard: about greed, consumption, and wastefulness (¶¶ 9, 55), and our "throwaway culture" (¶ 22), for instance. But these outward behaviors are themselves manifestation of deeper troubles. Among some people the underlying flaw is a hardness of heart, an "indifference or cruelty towards fellow creatures of this world": "We have only one heart, and the same wretchedness which leads us to mistreat an animal will not be long in showing itself in our relationships with other people" (¶ 92). More broadly, the problem is our lack of adequate virtue, our failure to act in ways that show gratitude for our gifts and that acknowledge our limitations. We are too prone to discount or ignore the common good as such; human ecology, we are told, is inseparable from it (¶ 156). And the "common" part extends to future generations (¶ 159). Instead, what we see is "the rise of a relativism which sees everything as irrelevant unless it serves one's own immediate interests" (¶ 122). Writ large it becomes a pernicious "culture of relativism," which includes "the mindset of those who say: Let us allow the invisible forces of the market to regulate the economy, and consider their impact on society and nature as collateral damage" (¶ 123). "When human beings place themselves at the centre, they give absolute priority to immediate convenience and all else becomes relative" (¶ 122).

To this view of moral relativism, of self-selected preferences, Pope Francis counters with the claim that the moral value residing in the natural order transcends us. On this point he connects with his overall normative vision tied to the ultimate flourishing of the entire community of planetary life, an ultimate goal or vision that he finds embedded in evolving nature as its inner, objective good. Given the moral values inherent in nature, given the inherent goodness of life and health, it is wrong for us as humans simply to seek "satisfaction of our own desires and immediate needs" (¶ 123). When "objective truth and universally valid principles are no longer upheld," then "the culture itself is corrupt" (¶ 123).

> The natural environment has been gravely damaged by our irresponsible behaviour. The social environment has also suffered damage. Both are ultimately due to the same evil: the notion that there are no indisputable truths to guide our lives, and hence human freedom is limitless. ¶ 6 (quoting Pope Benedict)

Knowing and thinking. This too-common tendency to elevate personal preferences over the common good is linked to flaws in the ways we reason. Too often we treat nature and other humans simply as tools to manipulate, as means to a self-chosen end, denying their inherent value as well as our common engagement in the larger, long-term community of life. Similarly, we ignore the mysteries and wonders of nature, denying the limits on what we know, while charging ahead with a purely scientific, technical vision of the world understood as physical stuff. "Many problems of today's world stem from the tendency, at times unconscious, to make the methods and aims of science and technology an epistemological paradigm which shapes the lives of individuals and the workings of society" (¶ 107). We end up "prizing technical thought over reality" (¶ 115), which is to say we interact mentally, not with the rich, morally infused real world, but instead with a science-based model of it constructed in our minds. The guiding "technocratic paradigm . . . exalts the concept of a subject who, using logical and rational procedures, progressively approaches and gains control over an external object" (¶ 106). It is not simply our use of nature that is the problem; it is the narrowness of our view and our ruthlessness in taking what we want. We "lay our hands on things, attempting to extract everything possible from them while frequently ignoring or forgetting the reality in front of us" (¶ 106).

The distortion here is worsened when knowledge is fragmented and action is guided by specialists who draw only upon learning contained within their specialties. "The specialization which belongs to technology makes it difficult to see the larger picture. The fragmentation of knowledge proves helpful for concrete applications, and yet it often leads to a loss of appreciation for the whole, for the relationships between things, and for the broader horizon, which then becomes irrelevant" (¶110). Too often, "technology . . . linked to business interests . . . proves incapable of seeing the mysterious network of relations between things and so sometimes solves one problem only to create others" (¶ 20).

Myths of power and progress. Guiding our onward degradation of nature is "the modern myth of unlimited material progress" (¶ 78), an "omnipresent technocratic paradigm" that links to "the cult of unlimited human power" (¶ 122). It is mythology that breeds "a false or superficial ecology which bolsters complacency and a cheerful recklessness [and] serves as a licence to carrying on with our present lifestyles and models of production and consumption" (¶ 59).

These various cultural ailments—our excessive anthropocentrism, our tendency to elevate personal preferences over the common good, our narrow, ruthless instrumental rationality, our embrace of myths of power and progress—characterize and shape the central institutions of society. They are reflected in the dominance of business elites, guided by this mentality, in nearly all spheres of social life, political as well as economic. "Our politics are subject to technology and finance" (¶ 54). Private property, similarly, has become disconnected from the common good and from the rightful duty of private owners to administer what they own "for the good of all" (¶¶ 93, 95).

Injustice and Population

These various root causes of ecological decline, Pope Francis reiterates, also underlie much social injustice toward individuals as such (particularly the poor) and among nations and regions of the globe. "We are faced not with two separate crises, one environmental and the other social, but rather with one complex crisis which is both social and environmental" (¶ 139). "Environmental deterioration and human and ethical degradation are closely linked" (¶ 56). The same mental frames, the same flawed cultural ills, are at work. "The mindset which leaves no room for sincere concern for the environment is the same mindset which lacks concern for the inclusion of the most vulnerable members of society" (¶ 196). Because the root causes overlap so much, efforts to deal with one set of problems need to address the other at the same time. "Everything is connected. Concern for the environment thus needs to be joined to a sincere love for our fellow human beings and an unwavering commitment to resolving the problems of society" (¶ 91).

If the present ecological crisis is one small sign of the ethical, cultural and spiritual crisis of modernity, we cannot presume to heal our relationship with nature and the environment without healing all fundamental human relationships. (¶ 119)

It is in the context of this linkage of environmental and social justice issues that Pope Francis in his encyclical brings in the matter of human population, viewed by many as a major cause of ecological decline. He does so in ways that challenge those who would lay heavy blame on a rising world population to consider what their reasoning about it too often entails and conceals. Is this claim, the pope asks, in reality mostly a way for the wealthy who play disproportionate roles in fostering degradation to push responsibility on to the poor? Is it an excuse also to deny the needs and possibilities for fundamental change in modern culture, in economic and political systems, and in the ways we all live?

The pope's consideration of population is largely confined to paragraph 50 (of 246). "Instead of resolving the problems of the poor and thinking how the world can be different," he begins, "some can only propose a reduction in the birth rate." Without denying that population can be problematic, particularly locally, Francis challenges the view. "To blame population growth instead of extreme and selective consumerism on the part of some, is one way of refusing to face the issues. It is an attempt to legitimize the present model of distribution, where a minority believes it has the right to consume in a way which can never be universalized." Still, he admits cautiously, "attention needs to be paid to imbalances in population density, on both national and global levels."

What to Do

Laudato Si', to reiterate, is best comprehended as a sharp, wide-ranging critique of modern culture. Its emphasis is on the root causes of our misuses of nature, not on the scientific evidence of those misuses. Normative factors are mostly used to identify and explain root causes and to present better cultural alternatives. They provide the raw materials for, but are not drawn tightly together into, a vision of good living in nature. For the pope, sound remedies for our ills need to mitigate or end these root causes. They need to foster significant changes to modern culture. We cannot put our hope in technology, existing or new. Nor can we expect individuals working within current cultural frames to remedy ills by making changes in their personal lives, by altering their individual preferences, by going green. What is needed, urgently, is "a bold cultural revolution" (¶ 114). Collectively we "need to slow down and look at reality in a different way," not overlooking the "positive and sustainable progress which has been made" but to go further and "recover the values and the great goals swept away by our unrestrained delusions of grandeur" (¶ 114). This means respecting "the intrinsic dignity of the world" and recovering our "true place in this world" (¶ 115); it means "liberation from the dominant technocratic paradigm" and embracing "another type of progress, one which is healthier, more human, more social, more integral" (¶ 112). "Put simply, it is a matter of redefining our notion of progress" (¶ 194).

As should be clear, this call for radical cultural change stands far from the claim today that new technology, widely used, can move us along. "Merely technical solutions run the risk of addressing symptoms and not the more serious underlying problems" (¶ 144). Unless we regain our compass, "any technical solution which science claims to offer will be powerless to solve the serious problems" (¶ 200). "To seek only a technical remedy to each environmental problem which comes up is to separate what is in reality interconnected and to mask the true and deepest problems of the global system" (¶ 111). It is simply wrong to believe, as some do, that "ecological problems will solve themselves simply with the application of new technology and without any need for ethical considerations or deep change" (¶ 60).

With language almost as strident the pope similarly challenges the much-heard call to promote of sustainable development as commonly understood as a guiding normative aim. "It is not enough to balance, in the medium term, the protection of nature with financial gain, or the preservation of the environment with progress. Halfway measures simply delay the inevitable disaster" (¶ 194). Too often, "talk of sustainable growth usually becomes a way of distracting attention and offering excuses. It absorbs the language and values of ecology into the categories of finance and technocracy, and the social and environmental responsibility of businesses often gets reduced to a series of marketing and image-enhancing measures" (¶ 194).

What is needed, along with a new understanding of progress, is a new, "distinctive way of looking at things, a way of thinking, policies, an educational programme, a lifestyle and a spirituality which together generate resistance to the assault on the technocratic paradigm" (¶ 111). This transformation includes, for individuals, personal renewals and "cultivating sound virtues" (¶ 211), particularly generosity, sharing, sacrifice, even asceticism, "moving gradually away from what I want to what God's world needs" (¶ 9, quoting Patriarch Bartholomew). At bottom, what is needed is a "profound interior conversion" (¶ 217). "When people become self-centered and self-enclosed, their greed increases. The emptier a person's heart is, the more he or she needs things to buy, own and consume. It becomes almost impossible to accept the limits imposed by reality. In this horizon, a genuine sense of the common good also disappears" (¶ 204).

Pope Francis is equally clear that much of the needed work, particularly institutional change and new laws, calls for collective action in politics and civil society. He invites and urges "a new and universal solidarity" (¶ 14):

> Self-improvement on the part of individuals will not by itself remedy the extremely complex situation facing our world today. Isolated individuals can lose their ability and freedom to escape the utilitarian mindset, and end up prey to an unethical consumerism bereft of social or ecological awareness. Social problems must be addressed by community networks and not simply by the sum of individual good deeds. (¶ 219)

The needed transformation, that is, must occur at the collective level as well as at the individual level. "The ecological conversion needed to bring about lasting change is also a community conversion" (¶ 219). Change of this type will not come easy, opposed as it is by the now-dominant elements of society.

> The alliance between the economy and technology ends up sidelining anything unrelated to its immediate interests. Consequently the most one can expect is superficial rhetoric, sporadic acts of philanthropy and perfunctory expressions of concern for the environment, whereas any genuine attempt by groups within society to introduce change is viewed as a nuisance based on romantic illusions or an obstacle to be circumvented. (¶ 54)

The path ahead, the pope makes clear, will require new collective rules for economic activities and land uses, new laws in individual countries and "urgently needed" "enforceable international agreements" (¶ 173). New laws, though, will not come about unless and until people insist on them, which is to say "a healthy politics is sorely needed, capable of reforming and coordinating institutions, promoting best practices and overcoming undue pressure and bureaucratic inertia." "In the absence of pressure from the public and from civic institutions, political authorities will always be reluctant to intervene, all the more when urgent needs must be met" (¶ 181). "What is needed is a new politics which is far-sighted and capable of a new, integral and interdisciplinary approach to handling the different aspects of the crisis." "A strategy for real change calls for rethinking processes in their entirety," for "question[ing] the logic which underlies present-day culture" (¶ 197). This means changing political business as usual. Given today's "culture of consumerism," it is too easy for government authorities "to rubber-stamp authorizations or to conceal information" (¶ 184). Worsening that problem is the reality that "lack of respect for the law is becoming more common"; laws can be well framed "yet remain a dead letter" (¶ 142). Further, "even when effective means of enforcement are present . . . the majority of the members of society must be adequately motivated to accept [the laws], and personally transformed to respond" (¶ 211). Cultural reform lies beneath and before all else.

It hardly needs saying that politics must gain independence from the economic forces that exert such control over it (¶ 189). This means, "once more, we need to reject the magical conception of the market, which would suggest that problems can be solved simply by an increase in the profits of companies and individuals. . . . Where profits alone count, there can be no thinking about the rhythms of nature, its phases of decay and regeneration, or the complexity of ecosystems" (¶ 190). Not everyone, the pope notes, needs to engage in the political arena, but enough people do. Those who do not should step forward, if and as possible, to support instead the "countless array of organizations which work to promote the common good and to defend the environment, whether natural or urban" (¶ 232).

That leaves, Pope Francis relates, the issue of education, understood not just in the classroom sense but as a public effort to instill new ways of thinking and talking about our common home and our place in it. Sound environmental education needs "to include a critique of the 'myths' of a modernity grounded in a utilitarian mindset (individualism, unlimited progress, competition, consumerism, the unregulated market)"; it needs, through the leadership of educators, to promote "an ethics of ecology," one that helps "people, through effective pedagogy, to grow in solidarity, responsibility and compassionate care" (¶ 210).

> If we want to bring about deep change, we need to realize that certain mindsets really do influence our behaviour. Our efforts at education will be inadequate and ineffectual unless we strive to promote a new way of thinking about human beings, life, society, and our relationship with nature. (¶ 215)

Looking Ahead

Comprehensive though it is, *Laudato Si'* leaves undeveloped several key points that need more clarity for readers to know adequately what it all means. A few of them are religious elements that arise within a document that is, in its ultimate stances, distinctly secular. Others involve points that Pope Francis perhaps intended to leave incomplete, as an inducement to readers to carry the issue further, or on which the pope in time might offer more commentary.

One religious element that poses concerns has to do with the pope's comments on ways of knowing, particularly means that reach beyond sentiment and respect for traditions to include insight gained by spiritual reflection. Readers have cause to be uneasy by this, given the ways prayer over the centuries has led people astray even as it has also brought gains. The concern is lessened because Pope Francis seems to limit prayer-based insights to issues of meaning, value, and purpose, not to scientific facts. By omission he rules out, as noted, attempts to dismiss our current ills by an appeal to God's inscrutable wisdom or to an imminent end of time. Throughout, his emphasis is on human responsibility and his sharp criticism is mixed with a remarkable hope that we can and will change our ways. So long as the pope's detailed view provides the overall frame, the worry about spiritual guidance seems small. Yet it does not disappear.

Similarly worrisome are the pope's comments about God's built-in purpose for us and the planet as a whole. As noted, the pope's teachings presuppose an ultimate vision of flourishing life on the planet. This is the built-in good or final cause of Creation, the mature stage toward which the community of life naturally trends. This vision might well be what the pope has in mind when instructing that the "ultimate destiny of the universe is in the fullness of God" and that "all creatures are moving forward with us and through us toward common point of arrival, which is God." Given our special intelligence and love, we humans "are called to lead all creatures back to their Creator" (¶ 83). We are left to wonder what this might mean, now or in time. In what ways and toward what end might we exercise our powers to lead? The pope's comments add a modest note of uncertainty and perhaps worry to a document that, overall, looks ahead rather clearly, toward a planetary community in which ecosystems retain fertility and resilience and life as a whole prospers.

These provocative comments about ultimate purpose give further reason to hope that the pope in time might say more about his overall normative vision. His normative principles, as noted, are many, having to do with the conservation of biodiversity, the promotion of ecosystem functioning, the protection of green space for all people, respect for individual rights to clean water, and more. Brought together the statements provide a fair enough outline of good land use. At the same time, they include sufficient vagueness—or more positively, they rely sufficiently upon broad principles—so that people in various settings might together decide for themselves how best to distinguish land use from abuse. The pope has put out the call for communities at all spatial scales to discuss the principles presented in his encyclical. He wants to stimulate engagement, not provide answers, even as he offers guidance on many relevant factors.

The pope, in sum, leaves us with something short of a clear vision of good land use to use as a normative end point. Again we might hope for more clarity in time. Yet the pope stands firm on the need for people to wrestle with the normative principles themselves. He calls for a process of deliberation to carry the normative thinking from broad principles to clearer understandings of how people might best inhabit their home lands. His role, that is, is to set the deliberative process off in a good direction, with sturdy intellectual, moral, and spiritual tools, and then to serve as cheerleader to encourage movement in a sound direction.

Another soft point in *Laudato Si'* arises from the pope's insistence that our misuses of nature—including our disrespect for other creatures and species and for future generations—cannot be addressed without also confronting the challenges facing the world's poor, including the scandalous maldistribution of income and wealth among regions and nations. The problems, he rightly points out, are interlinked, closely so in terms of their underlying cultural causes. Means to address them thus ought to work in tandem. All of this is clear enough.

What's missing given today's tendencies is express comment on whether efforts to address one of the dual challenges might rightly employ means that worsen the other challenge. Can we, to use the obvious example, promote economic development projects that foster further degradation even as they do aid the poorest? Can we lump the two concerns together and charge ahead so long as we predict the overall consequences on balance will bring improvement? Or are we constrained to use only means that do not worsen either problem?

As noted, Francis challenges those who would promote a vision of sustainable development aimed too much at adding profits to international developers and banks (¶ 194). "It is not enough," he instructs, "to balance, in the medium term, the protection of nature with financial gain, or the preservation of the environment with progress. Halfway measures simply delay the inevitable disaster" (¶ 194). Further, when speaking of poverty he routinely points a finger at overconsumption by the rich, including the wealthy elites of countries in which poverty is widespread.

We should be particularly indignant at the enormous inequalities in our midst, whereby we continue to tolerate some considering themselves more worthy than others. We fail to see that some are mired in desperate and degrading poverty, with no way out, while others have not the faintest idea of what to do with their possessions, vainly showing off their supposed superiority and leaving behind them so much waste. (¶ 90)

The pope at no point endorses further degradation of nature or disrespect toward its many values as an acceptable price of aiding the poor. The issue is important enough to want more clarity, and yet the encyclical as written, omissions included, is perhaps meant to be clear. So much economic development, intended to aid the poor, has no such good result, we are told (¶ 194). Given that reality, and given the vast opportunities to aid the poor by altering systems of income allocation, there is little need to cut further into nature. Enough change can come by cutting the incomes and wealth of the rich.

Adding to the vast numbers of poor people, of course, is the still-high birthrate among many poor populations, adding not just more mouths to feed but new millions who will strive to move up the consumption ladder. One can sense in *Laudato Si'* that the pope wanted to flag the issue and make a few key points but not to address it, in the encyclical, with any fullness. As noted, he challenges those who blame the reproducing poor for global problems without acknowledging their own contributions and changing their own ways of life. It is a strong and valuable claim, and rightly stressed. But the issue of overpopulation nonetheless remains. The key issue for the church is its stance on contraception. Much of the world ignores the church's opposition to it. But that opposition, picked up by other Christian groups, nonetheless shapes public policies and makes access to contraception, or even knowledge about it, harder for many women while giving cause for others to criticize women who want access. In *Laudato Si'* the pope stresses the diffusion of moral value among all life forms, as species and communities and as individual creatures. At what point does the expansion of the human population infringe upon that contrasting moral value; at what point does it contribute too much to the social injustice that he decries? On the issue of population *Laudato Si'* is plainly inadequate.

Finally, there are the pope's various alluring comments about the need to rethink the institution of private property. The pope hardly rejects private ownership. Yet he implies that we err greatly in thinking that such property exists chiefly to benefit owners and that rights-holders can use what they own simply to achieve private gains. In language likely unfamiliar to many readers he insists that all forms of property are infused with "social purpose" and subordinate to "the universal destination of goods," which is to say that private things may only be used by owners in ways consistent with the common good (¶ 93). Quoting Pope (now Saint) John Paul II, Francis instructs that "there is always a social mortgage on all private property, in order that goods may serve the general purpose that God gave them" (¶ 93). As for private rights in nature itself—that is, in the common home entrusted to us—we must see that it is "essentially a shared inheritance, whose fruits are meant to benefit everyone" (¶ 93).

These and related comments together pose a stern challenge to prevailing ideas of private ownership, particularly within the United States and other developed countries. Private property is a morally complex arrangement, capable of yielding communal gains along with private profit but capable also of empowering owners to exploit nonowners, to divert unfair shares of income, to manipulate political organs, and otherwise to bring on the many ills that Pope Francis laments. His comments point to a possible new understanding of private property, one that is more ecologically and socially sensitive while still retaining the important benefits that a well-designed scheme of ownership can generate. So pervasive is private ownership, so problematic is it becoming in many nations, that the topic seems worthy of separate treatment.

FIVE

Taking Property Seriously

The private ownership of land has become a considerable problem for many people who care about nature and who worry that we are not living responsibly in it and on it. Too often, it seems, landowners act in ways that are not ecologically sound. Too often they think only of themselves and the short run and fail to consider the larger community of life and future generations. And private property—that is, the legal rights that owners possess—seems to shield their land uses from meaningful challenge. For environmental reform efforts, it is a significant obstacle on the path ahead, routinely encountered whenever new land-use limits are proposed.

These worries about private land use build upon a long history of varied concerns about private property as a social and economic institution. Generations ago, the predominant worry was different; it had to do with the ways landowners dominated the lives of their tenants and other people who used their lands. Control over nature meant (and still means) control over the people dependent upon that nature for survival. This issue of power remains alive in much of the world, even as it has been eclipsed in developed countries by environmental concerns. In the 1930s the concern of the day had to do with the property rights of corporations and the economic and social havoc brought on, critics said, by their misuse.[1]

Matched with these various worries about private property, though, has been a deep-rooted, widely shared respect for this familiar institution. Private ownership is, in fact, a valuable if not indispensable component of the social order and of economic flourishing. As a system of orchestrating land uses it yields substantial benefits, not just for owners but for society as a whole. The pertinent question, then, is this: Can we reform the institution so as to keep its good consequences while curtailing or eliminating the bad ones? Can we revise the rules of ownership so as to promote ecologically sound land uses without undercutting property's many longstanding benefits? Or, alternatively, might our choices be starker ones: to seek good land use through some form of collective ownership or instead to retain private property and either learn to tolerate its ecological costs or start paying landowners to change their degrading ways?

Ironically, one of the challenges of addressing this important issue is the very pervasiveness and familiarity of private land ownership. Private property is all around us. We know what it entails, or so we think. We know its fundamentals, at least well enough to debate the means or tools we might use to get landowners to act better.

In truth, we do pretty much know what property means, or at least what it means today in the particular legal jurisdictions where we live. What we know far less well are the fundamentals of the institution. Hurrying to find solutions to land-use ills, we rarely take time to consider where property comes from, why it exists, and how it operates. We do not study its history and many present-day manifestations to understand its flexibility and to trace its evolution over time. In short, we fail to take private property seriously as an object of study and reflection. It is a costly failure, because we commonly assume, wrongly, that property is less flexible than it really is and that it inevitably, and somehow legitimately, vests landowners with legal rights to behave in ways a community might dislike.

This inquiry into the institution focuses, appropriately, on the private ownership of nature itself, of land, water, timber, and the like. Private rights in nature differ from property rights in other things—in chairs, microphones, computer software systems, or drug patents. People did not create nature; that is one significant difference. Nature is relatively finite (another difference) and it is essential for human welfare (yet another). If all of Walt Disney's movies suddenly disappeared, human life would go on, for better or worse. If all of the earth's topsoil washed away, human life would pretty much end. We live in nature and its parts are interconnected. Our long-term welfare depends upon its healthy functioning. Owning nature is different.

Private property is a human, social arrangement that can take and has taken a wide variety of shapes and forms over time, in terms of what can be owned and what it means to own. It is also a morally problematic arrangement, given its various effects on people, and it can be justified morally only to the extent it promotes the common good. The elements of ownership have evolved considerably over time, even in their fundamentals. And they ought to continue evolving to reflect changed circumstances, public values, and ecological understandings. How, though, can this evolution best take place so that private property continues to work well overall? For environmental reformers this is the basic challenge. To take it up one needs to look at the institution as a whole, seeing what it is, how it works, and how new ownership norms might challenge cultural ills and foster better land uses.

One of the grave deficiencies of the modern environmental movement has been its failure to engage this issue. Typically green groups have stood by quietly while opponents have raised high the property-rights flag, proclaiming a vision of property rights distinctly hostile to environmental concerns. They have said little while opponents claimed the high ground as defenders of private property and portrayed environmental groups as its attackers. The usual response, rather than to challenge the vision, is to argue with facts and figures that the invasions of property rights are worth the cost.

A much different approach would have been, and remains today, far wiser. Property law need not give landowners the power to act contrary to human flourishing and the common good. Further, well-structured property rights can often aid the cause of environmental reform. Equipped with a new, well-considered understanding of private rights, environmental reformers need to present themselves as proponents and defenders of private property. What they need to defend, openly and clearly, is responsible land ownership. What they need to resist, and to pin on their opponents, is irresponsible land ownership.

The main elements of a new vision of ownership are sketched in the final part of this chapter, which on several points looks ahead to discussion in later chapters. It is a reform platform that recognizes, as it must, that a debate over the meaning of private property immediately engages and challenges key cultural values. Modern culture needs to evolve if a new, ecological understanding of ownership is to take root. At the same time, culture change won't likely yield better landscapes if it does not translate into better ownership norms.

The Fundamentals of Private Property

Private property, as noted, is very much a human-created, social institution. It deals with relationships among people with respect to nature and its parts. Lawmaking communities over time have differed considerably in terms of what parts of nature a person can own, how owners can use nature, and how the entitlements of one owner fit together with the entitlements of other owners and the surrounding community. How are property rights acquired and how long do they last? Answers among societies have varied. Must owners use what they own or risk forfeiture of their rights (a use-it-or-lose-it rule)? What rights do landowners have to exclude outsiders? Perhaps most basically, is landownership in the contemporary Western sense even authorized by a particular legal system or does the system instead rely on more narrowly tailored use rights?

A corollary of this social reality is that private property has no paradigmatic form—there is no Platonic form or ideal—nor is there any such thing as full or complete ownership to serve as an ideal or benchmark. Landowners can have more or fewer rights and the law can define their rights in varied ways. Put otherwise, to say that John owns Greenacre really means very little unless we know what ownership means at a given time and place. The rights of one landowner are necessarily intertwined with the rights of others. If one landowner can engage in noisy, polluting activities, then neighboring owners may have to put up with the disruption. Instead, if neighboring owners can successfully complain about the disruption and halt it—exercising their rights to be free of disruption—then the landowner who wants to make noise faces limits on her rights. On this issue and countless others, lawmakers have to draw lines and make choices. Either they allow noisy, polluting activities or they protect the quiet enjoyment of neighbors. They cannot do both, which means private rights, however defined, are necessarily qualified.

In truth, private property has almost no core content that is recognized in all times and all places. Nineteenth-century historians such as Sir Henry Maine thought that property rights schemes naturally evolved over time, heading toward something like the common law regime in place in nineteenth-century Britain.[2] Such evolutionary meta-theories have long since lost favor, particularly among serious historians. Indeed, many tribal societies had remarkably sophisticated property arrangements, often arrangements that featured clearly defined *use* rights and that did not aggregate or bundle these specific rights into anything like the allodial, fee-simple ideal so exalted by Anglo-American lawmakers.

In many debates about private property in the United States the common narrative of conflict has long pitted an individual landowner doing battle with a regulatory state. Anecdotal stories are put forth, typically with only two characters: the landowner on one side, the state on the other. In fact, though, government's role is usually different than this and certainly more complex. More often, a real-life dispute pits one landowner against a neighboring landowner, or one landowner against the aggregate of landowners acting as a community. In such cases, the law is called upon to resolve the conflicts among the various owners with property rights existing on both sides of the dispute. There is no pro-property position for the law to take; instead, we have conflicting or contrasting views about the rights landowners ought to possess. Accordingly, the question in such disputes is not whether and to what extent we should protect private property. It is, rather, how we ought to define private property. Should landowner A have the right to drain water from his land if it would harm landowners B through Z? Should landowner B be able to build a tall building that blocks the sunlight or air coming to neighbor C or adds congestion to streets that serve all other landowners?

This complexity is easily lost when we use simple anecdotes involving a single landowner pitted in battle against the state. Our stories need far more characters to have value: other landowners, landless people, other life forms, perhaps future generations. It needs emphasizing that private property does not exist independently of law and state power. Property is a product of law; as Jeremy Bentham said nearly two centuries ago, take away the law and property disappears.[3] And when we talk about law creating and recreating property over time, we necessarily mean all applicable laws. Laws do not come in two types: those defining private rights and those regulating the use of rights. Laws are laws; they differ only in terms of the lawmaker and their applicability. The rights of ownership arise out of the intersection of all of them. Finally, we must not forget that the power exercised by landowners is public, governmental power, not some sort of private power that exists in another realm. A landowner's power stems from his ability to call upon police and courts to restrain other people, maybe even imprison them. This is the same coercive, sovereign power wielded by police in criminal justice actions, and it should be equally subject to scrutiny.

Simply by the way it operates, private property in nature is a morally problematic arrangement. The moral problem arises because private property, created by law, gives an owner power to impair the liberties and flourishing of other people. When the law recognizes Albert as the owner of Greenacre, it gives Albert the power to limit uses of Greenacre by everyone else. When Albert puts up a "no trespassing" sign and wandering hunter Xavier gets arrested for trespass, this is, plainly, a moral issue that calls for justification. How can we justify arresting hunter Xavier, perhaps even imprisoning him? We cannot justify the arrest by saying that Xavier violated Albert's property rights, because the property rights themselves are what we are trying to justify.

Consider, as another instance, the case of farmer Bill, who tills the land in the spring, plants crops, tends them, harvests them, takes them to market; in short, he does everything. Then, on market day, along comes Carol who says, "Give me half of your earnings." How can this be right? How can Carol, who did no work, demand half of the revenues from the land? Does not this violate the longstanding labor-producerist theory of moral deserts?

Of course it does. And again, we do not avoid the moral quandary simply by attaching to Carol the label of landowner. Why is it that a landowner who does nothing gets half the land's produce? The familiarity of the institution hardly makes it morally right. More justification is needed.

It helps to go back to the beginning of things. Mythology aside, private property does not arise when John seizes a tract of land and announces to the world, "This is mine." It arises only later, when other people agree to recognize John's rights and then make some arrangement to protect these rights. But why should they do so? Why should they let John seize part of the common pool of land and claim it exclusively? They would do so, presumably, only if the arrangement benefits them, only if the recognition of such property rights benefits not just the owner but other people, including nonowners.

This truism about the moral complexity of private property was at one time widely understood, but it has been largely forgotten, particularly in the United States. These days American writers tend to think that property exists chiefly to benefit the owner, and that property's moral issue arises when government tries to restrict what an owner can do. It is the restriction of property rights that is morally problematic, according to many people today. This perspective greatly distorts the moral landscape. When John Locke, Jeremy Bentham, and others wrote about the philosophy of private property, they viewed as their challenge the task of justifying private property itself. They knew how property worked. They knew that a valid justification had to explain why nonowners should respect the rights of owners, and why property should otherwise authorize owners to act in ways that cause harm. Today's writers largely assume property's moral legitimacy, thus avoiding the fundamental issue.

What ought to be clear is that we cannot justify private property by looking only to the ways it benefits the owner. Of course Carol is better off if she gets half of the farm's income, just as Bill is worse off. It is the burden on Bill that needs explanation. To justify such burdens—to justify private property generally—we have to explain why it benefits other people, owners and nonowners alike. The Anglo-American writer Thomas Paine was one of many who directly engaged this moral challenge.[4] Like others of the Enlightenment era he believed that the earth was owned by people in common, and that each person upon birth gained an equal share of the earth, an equal right to use it along with everyone else. Private property restricted this right. How could this be legitimate?

We could try to resolve this quandary by arguing that private property is based on some inherent individual right, and that, when the state arrests trespassers, or enforces landowner Carol's claim to half the farm's yield, it is merely protecting or enforcing an individual right, perhaps based on natural law. This argument, though, hardly withstands analysis. Where did such a right come from (a severe problem of all "natural rights" arguments)? And how did it take precedence over the liberties of other people (also, of course, natural rights)? To make this argument one would need to assume some Platonic ideal in property, arising out of nature itself. The claim makes no sense given the historical record. Even leading libertarian works such as Robert Nozick's *Anarchy, State and Utopia* offer no proposals for justifying private rights in land under real-world conditions of scarcity. Ultimately, like others, Nozick had to resort to a moral justification for private ownership based on the only sensible basis: overall social welfare.[5]

In the United States, a lot of today's ideas about private property date from the nineteenth century. That was a time when the dominant sentiment in the country favored westward expansion and the rapid development and exploitation of the continent's resources. Americans wanted to settle and tame land, and, in the process, make lots of money. Given this policy orientation, it made sense simply to unleash people on the land and pretty much let them do as they pleased, trusting market forces and self-interest to provide traction and guidance. That was what citizens wanted, and American lawmakers shaped property law to help do it.[6]

It is worth stressing that American property law required rather substantial change in order to empower individual owners to exploit nature fully, to dam rivers, divert water flows, clearcut forests, reshape terrains, drain marshes, and the like. The private property system that entered the nineteenth century was apparently quite protective of settled, quiet, agrarian land uses. It was not legally permissible for an owner of land to use it in a way that harmed existing surrounding land uses. To unleash industrialization, this basic legal orientation needed rewriting. Landowners needed to gain far greater powers to use their lands intensively without much regard for the resulting ill effects on other landowners. Necessarily this meant that their neighboring landowners had to relinquish some of their former powers to halt activities by neighbors that interfered with their agrarian land uses. What happened is that the right to use land intensively became more expansive while the right to halt interferences—the right to quiet enjoyment—was curtailed. Property rights did not increase overall; they were simply reconfigured. That was the dominant evolutionary trajectory in the nineteenth-century United States, appearing in many particular areas of property and natural resources law.

A big problem today in the United States if not elsewhere is that we do not have a good sense of private property's history. We do not realize how extensively prior generations reframed private rights to serve what they viewed as the public interest in their day. Lacking this historical perspective, we seem to view certain property arrangements as timeless, when in fact they are anything but.

Private property has typically been an institution on the move, even if changes happen slowly enough that many people miss them. Some changes are overt: a new statute bans construction in a floodplain or on a wetland, thereby redefining the right to develop. Sometimes changes are seemingly more modest, as when a court recalibrates the water uses that are deemed "reasonable" or decides that a coal-mining company needs to pay for the surface damages that it causes. Over the long run, however, even small changes add up. In many countries the right of landowners to develop has gone down appreciably, particularly the right to alter lands that are ecologically sensitive. And more changes loom on the horizon. Should farmers have the right to till lands in ways that cause net soil loss? Should they have the legal right to apply farm chemicals that run off into rivers and lakes? Should landowners face limits on their ability to alter vital wildlife habitat or to disrupt natural water flows? The law's answers can change, and likely will.

One particular idea that has arisen in the United States is the notion that land ownership inherently includes something close to a complete right to exclude other people from the land.[7] This is now viewed by many as a core element of ownership that is or ought to be timeless. This claim is surprising because trespass law in early America—up to and in many places well beyond the Civil War of the mid-nineteenth century—was quite different. Landowners in the United States did not have full rights to exclude. They could typically exclude other people only from lands that they enclosed with expensive, sturdy fences. This meant that the vast bulk of the countryside was open to public use, without regard for landowner wishes. It was open for public hunting, livestock grazing, firewood gathering, and so on.

The right to exclude is really no different from any other component of landowner rights, which is to say that it is morally legitimate only to the extent we can justify the right by reference to the common good. Perhaps it made sense in the late nineteenth century to expand the rights of landowners to exclude, closing what had been a vast domain open to public use by rich and poor alike. But we should view the trespass laws then put into place as no more durable than the trespass laws that they displaced. Property laws need to evolve over time to align with the overall public interest. As that interest shifts, due to new circumstances, knowledge, or values, then the law ought to shift along with it.

An increasingly fragmented, self-centered American culture has seen fit to allow landowners to keep the public away, even when public uses do not interfere with anything a landowner is doing. As this development has reached its final stages, America's legal ancestor, Great Britain, has been moving in just the opposite direction, taking guidance from various Scandinavian countries. Britain has expanded its extensive footpath system by opening up vast rural lands not being tilled to public recreational uses, proclaiming a public right to roam.[8] Landowners, the British admit, should not be disrupted in their chosen land uses. But why should private owners be able to lock up lands—parts of the common home of all British people—when modest public uses would not be disruptive? To give landowners an unlimited legal right to exclude is to vest them with powers that diminish the common good. It is to curtail the liberties of public hikers and others for no morally sound reason.

Two Recurring Issues

To recapitulate: private property is very much a human creation that lawmakers at various times and places have shaped to meet their needs. This shaping and reshaping is not just appropriate but morally necessary given that private ownership is morally problematic. Private ownership entails the use of state power, put at the beck and call of private owners, to control parts of nature and thus to influence if not dominate the lives of people whose survival (or at least flourishing) depends on that nature. Such arrangements need justifying, and they are justifiable only to the extent they promote the common good through widely shared benefits. As circumstances, values, and knowledge shift, the common good is likely to shift along with them. That creates, in turn, a need for shifts in property law so the institution does not lose its moral legitimacy, so that it does not authorize landowners to wield power in ways that are, overall, harmful to the community at large.

What is the common good? What, then, is the common good when it comes to private property rights in nature, and how might we define it? To answer we need to start with private property and consider its various benefits, seeking to the extent possible to maximize the overall gains society obtains from it consistent with considerations of social equity. Then we can take up the issue of ecological degradation or, more positively, the issue of how we ought to live in and on nature. How might we best express our overall environmental goal? When this goal is clearly expressed, it then becomes more possible to recalibrate the elements of private ownership to help achieve it.

Theorists over the centuries have seen a variety of possible benefits that can come from a well-defined system of private ownership.[9] The most obvious benefit to people today is the ability of private ownership to promote overall economic development. Property can encourage and facilitate private efforts to mix labor with land and create things of value. It can encourage farmers to plant crops in the spring and harvest them in the fall. It can encourage people to build homes, stores, and factories, to sink mines, to build irrigation ditches and the like, in ways that grow the economy and yield widespread public benefits. Private property is, in fact, a useful if not essential tool in promoting growth. The evidence here is overwhelming when we compare cultures with secure private rights with those that lack such rights.

When the United States was founded in the late eighteenth century a more common argument about private property took center stage. The dominant claim then was that the widespread ownership of land diffused power and gave owners critical stakes in civil society.[10] These effects in turn helped strengthen and sustain popular governments and, in the view of many, were essential to the long-term maintenance of democracy. Writers spread over the next century elaborated this political justification. They contended that democracy could succeed only if governments made land available to as many people as possible, on easy terms, so that private ownership would become widespread. Though this rationale is less often heard today it is by no means unimportant. People with long-term economic stakes in a community are more likely to defend it and promote its welfare.

A third way private property can benefit society is that it enables people to craft spheres of personal privacy, places to which they can retreat from public gaze and engage in self-chosen means of development and living. This justification seems to focus on the way property benefits the individual owner, but the point is better expressed more broadly. Society as a whole becomes better and stronger when it facilitates individual growth and expression. Public tensions can dissipate, misfits can be handled and experiments tried, when private land offers places of retreat.

These three arguments form the central justification for private property rights in nature. Private property, well crafted, can benefit the common good by promoting economic enterprise and development, by adding ballast to the civil state, and by protecting privacy. Whether private property delivers these benefits depends on how it is structured and how it operates. Well crafted, the institute can provide vast benefits. When poorly crafted the benefits go down. As the inquiry so far suggests, there are two quite different dangers here when property is not well crafted, when the powers vested in landowners are not calibrated well with the common good in mind. On one hand, the institution can become morally problematic. It can give landowners the ability to use what they own to control other people, or extract wealth from them, in ways that are not morally legitimate. It can enable private owners to call upon the power of the state—its police, courts, even prisons—to restrain the liberties of other people under circumstances that are not morally legitimate. As for the second danger, a poorly crafted property system can simply fail to deliver the goods. It can fail to generate for society the full benefits that are possible from it in terms of economic flourishing, political stability, and the enhancement of individual privacy and liberty. Then we all lose.

Plainly, the common good is fostered in the long run if humans can find enduring ways to live in nature, ways that maintain nature's productivity and are thus sustainable. The more clearly that we articulate this aspect of the common good, this much-needed environmental goal, the better lawmakers can recalibrate private ownership to promote the overall common good. How, then, should we be living in nature? Put otherwise, what is the rightful role of humans in the web of life, given its interconnection and interdependencies? Or again, if we inhabited a landscape with the intent of residing on it forever, generation upon generation, how would we live in relation to the soils, waters, plants, animals, and processes that compose the natural order?

The questions are by no means easy to answer. Aldo Leopold had his answer, as we saw in chapter 1, brought together in his normative vision of land health. Wendell Berry among others, we have noted, has come up with a similar ideal, though less overtly grounded than Leopold's in ecological functioning. A well-constructed normative goal requires no small amount of careful thought, more thought than the environmental movement overall has given it. This thought needs to consider, among other issues, questions about moral value, about rights and duties, and about ethics and the good life. Where does moral value reside in the universe? In humans alone; in other life forms as well; in species or communities as such; in future generations? Similarly, if we were to live as virtuous people, what virtues would we embrace, and how might we translate those virtues into daily life?

In international circles the normative vision of *sustainable development* has held considerable influence and still does, despite continued criticism that the concept lacks much meaning. In the United States the tendency has been to use the related ideal of *sustainability*, again notwithstanding contentions, previously noted, that the term is rather empty of content and thus means all things to all people. More recently, the notion of *ecosystem services* has come to the fore.[11] This conservation vision is what might be termed a functionalist approach in that it seeks to sustain the ecological functioning of natural systems rather than trying to maintain all of their biological parts.

This new writing on ecosystem services tends to overlook claims of moral value in nonhuman life forms. Also, it only vaguely engages with claims that we have felt moral obligations to future generations of humans.

The value of a well-articulated conservation goal is that it adds substance and clarity to understandings of the common good. If we could see clearly what good land use entailed, we could judge existing land uses in relation to them and identify land uses in need of change. A reasonable starting point would be to call into question elements of property law that empower landowners to use land in ways that are bad when judged by standards of good land use. Of course property norms cannot simply change overnight, and the benefits of property are diminished when landowners have too little protection for existing land uses. Still, property laws, as explained, have slowly evolved for centuries, and the desire of landowners for stability and predictability are properly weighed against the costs of allowing harmful land uses to continue. A more full inquiry would likely show that landowners do need substantial protection in existing land uses, at least to the extent necessary to recoup investments in improvements to land. They need far less protection for their hopes to engage in new land uses in the future, including new development. Restrictions on future development are far less costly than interferences with existing activities in terms of the ability of private property to promote sound economic growth.[12]

These comments hardly go beyond introducing this issue of the common good in the property law context, the issue of how ownership systems might promote human flourishing and the ecological health of all landscapes. Merely to put the topic on the table, however, is to advance the inquiry. Merely to talk about the elements of private property in the context of the common good and human flourishing, keeping property's core values in mind as well as conservation ideals, is to supply a framework for more detailed talks about how new rules of property ownership might best serve long-term needs.

Property and liberty. Before moving on it seems useful to take up an obvious complaint or worry. With all of this talk about the common good, whatever happened to private property as an individual right? What happened to the notion that property protects individual liberty—that the right to own property is, some would say, the keystone individual liberty upon which all other liberties depend? Surely there is truth to this, truth the environmental movement, in challenging existing property norms, needs to take seriously.

The answer, as often, is mixed. The claim that property promotes individual liberty is wrong in that private rights in nature do not exist primarily to promote or protect individual liberties of any sort. They promote first and foremost the common good. One cannot start with an abstract notion of individual liberty and get to a vision of the common good, or least no one ever has. On the other hand, the common good is often served by a property system that does create and enhance individual liberties, that does give people legal rights to interact with nature, shaping the land and making it yield. That is, the law can often promote the common good by creating and protecting individual liberties. When that happens, we can rightfully say that property and liberty are linked, and that property protects individual liberty. But the connection is indirect and secondary. We protect individual liberty in land when and to the extent doing so promotes the common good, and the types of liberty that get protected, when they clash with other liberties, should be those that best serve the common good.

To arrive to this point is to turn upside down the rhetoric that landowners typically use. A landowner says: I ought to have the right to do x and y on my land. Or she says, you have violated my rights by restricting my ability to do x and y, and you ought to pay. Rights rhetoric of this sort, of course, has a long history. People assert rights in all manner of settings in which the claimed rights do not in fact exist. What such rhetoric means is that the advocate wants the law to create or recognize such a right. Such claims are fair enough, as forms of public argument. As we evaluate these claims, though, we need to rephrase them. We need to hear the landowner as making instead this claim: I should have the right to do x and y on my land because, when the law allows me to do that, when it allows me and other landowners similarly situated to do x and y, then the *public* will benefit. That is how we ought to hear the claim and then assess it. In the case of each contested right we need to ask: does the recognition of this right benefit society generally?

The Elements of Ownership

What kinds of changes to property norms, then, might the environmental reform effort propose to help foster new visions of the common good? How might it, in particular, propose reframing landowner rights so as to promote good land use at the landscape scale, or at least weed out land uses that are the most ecologically degrading?[13]

This topic is no less complex than others already raised, but it is somewhat easier to list the main elements of ownership that should be up for reconsideration and adjustment.

Redefining land-use harm. For starters, property law has always included, as a guiding principle, the sense that owners are obligated to use what they own in ways that do not cause material harm, either to other owners or to surrounding community. This is a basic do-no-harm principle, sometimes expressed in Latin, sometimes by using the French term seized upon centuries ago, nuisance. The basic idea is important yet obviously vague. What actions are harmful? The answer is that actions are harmful when lawmakers at any given time and place view them as such, which is to say that harm is an evolving social concept that lawmakers adjust to reflect prevailing conditions and values. For those who think property rights ought to be timeless this interpretation can be unsettling. But property rights have never been timeless, nor should they for moral reasons notwithstanding the benefits that come from stability and predictability. Today's definitions of harm were inherited from earlier generations, who for the most part felt free to redefine the term as they saw fit. If property has any foundational background principle, it is the principle that lawmakers are free to redefine harm, generation upon generation.

This idea of harm is certainly flexible enough to take into account one of today's dominant forms of harm, the harm that comes from a land use that is perhaps appropriate in isolation or when a few landowners engage in it but that becomes harmful when too many landowners do. These could be termed carrying-capacity harms in that they involve human activities that exceed a landscape's carrying capacity. Plainly, diversions of water from a river or aquifer can exceed carrying capacities, even if each water use in isolation seems reasonable. Landscapes can slide down when too much land cover is changed, when too much land is regularly tilled, when hydrologic flows (e.g., drainage) are materially altered, and in various other circumstances. In such settings, we could define harm in ways that, in effect, expect landowners to do their fair share in keeping their landscapes healthy. The "fair share" ideal is one that resonates widely among people across the political spectrum, in rural communities in particular. It is language environmental reformers could use regularly.

Tailoring private rights to nature. New definitions of land-use harm could be part of a larger initiative to redefine the property rights of a landowner to take into account the natural features of the land that is owned. The basic idea here is that a landowner's ability to use and reshape nature should derive in important part from the physical features of the particular land, with land uses limited to those consistent with the maintenance of basic ecological processes. A developer with a project in mind would thus need to find land that was ecologically, as well as economically and socially, well suited for it. The principle has many possible applications. It represents a significant deviation for the Anglo-American property thought that emerged in the mid- to late nineteenth century. Then, formalist thought was on the rise, and scholars (quite conservative by and large) increasingly talked about property rights in the abstract—as rights that attached to the ownership of hypothetical land parcels such as Greenacre or Blackacre and that paid no attention to physical features.

One of the prevailing myths surrounding private property outside urban areas—away from places with extensive zoning—is the idea that the rights of landowners ought to be essentially the same even though land parcels differ widely in terms of terrain, soils, drainage, vegetation, and resident wildlife. But they need not be, not when acceptable land use is so dependent on such natural factors, not when land uses need to reflect natural differences if they are to support the common good. One cultural element at work in this thinking is the valuable ideal of equality. Landowners want to be treated equally. But equality means treating owners as people without regard for morally irrelevant differences among them such as race or sex. It does not mean giving equal treatment to landowners when the lands they own differ significantly. Land-use rules that reflect nature treat land parcels differently, not landowners.

Reviving and protecting public property rights. A reformulated scheme of property law would likely pay more attention also to those parts of nature that are ultimately and by long tradition owned by the people collectively. In the United States the people collectively own all water flows and wildlife, even when on private land, as the law has pronounced for many generations. Wildlife remains public property until lawfully captured; water remains publicly owned although private parties (landowners and others) can acquire specific use rights in the water flows. A special legal status also attaches to land beneath navigable waters, which includes not just tidal waters but all inland waterways that are navigable in fact. By longstanding law these submerged lands, even when private owners hold title, are subject to public rights that take precedence over the rights of the private title holders. Private owners under the public trust doctrine hold subject to public use rights, including public rights to travel, fish, and swim in the waters without need of landowner permission.[14]

These special public rights could become far more important legally than they have been. If the public owns wildlife, even on private land, then presumably it has a legitimate claim that land uses make room for that wildlife. If water is publicly owned, wherever located, the public might properly insist that private actors avoid disrupting water flows in ways that are ecologically degrading. In many settings, a wide range of questionable land uses (e.g., excessive land cover removal, tillage, drainage, chemical usage) end up degrading public waters and reducing public uses of public waterways. Lawmakers might appropriately redefine private rights in land so as to limit such activities.

Property myths and the common good. For a vivid illustration how the mythology of private property causes heavy burdens today we can look to the ongoing struggles in California with water shortages (as the problem is termed, itself part of the problem). All water, as noted, is owned by the people collectively. What the holders of water rights possess are carefully defined and limited use rights, which owners can exercise within the prescribed limits. From the beginning, water law allowed rights-holders to use water only in ways that were socially beneficial—the "beneficial use" requirement, as the law terms it. In California, according to a provision of the state constitution (art. XIV, sec. 3), water uses must at all times also qualify as reasonable, which is to say reasonable from the perspective of the needs, understandings, and values of the people. And the legal limit is clear: unreasonable uses are not authorized and must halt.[15]

The problem in California, and to varying degrees elsewhere, is that massive flows of water are devoted to low-valued agriculture uses, to growing pasture grasses, hay crops, and corn and cotton in desert areas. The resulting commodities are not at all in short supply in the nation as a whole, and they can easily be produced elsewhere without irrigation. Indeed, the nation's central agricultural problem since World War I has been the challenge of handling overproduction. A key method to do that today is the program to mandate and subsidize uses of farm crops as biofuels (ethanol mostly), even when the fuel energy produced by the crops does little more than offset the energy needed to grow the crops, harvest and dry them, and then convert them into biofuels. The ethanol program overall does little to improve air quality and hardly more to reduce uses of fossil fuel. The program is promoted by farm interests, and it is promoted simply as a way to get rid of the excess commodities and to increase income for farmers and their industrial suppliers. Midwestern corn going to ethanol could be used instead to replace corn grown by irrigation. Other lands devoted to ethanol crops could readily produce the cotton and hay crops now blooming in desert areas, fully meeting national needs.

Viewed at the large scale, the vast bulk of irrigation water used in California—perhaps as much as half of all the water used in the state— yields little economic value. And that value comes at grave ecological cost given the direct harms of irrigation to soils, waterways, and aquatic life. As bad, it comes at huge economic expense to other water users who must find expensive alternatives, most perniciously desalinization. When all relevant factors and options are considered, it is hardly plausible to claim that low-valued irrigation—in places such as the Imperial Valley—is at all reasonable. If it is not reasonable then landowners have no legal right, no private property right, to engage in it. They are thus diverting and consuming publicly owned water without authorization. And they can be told to halt, now, without compensation.

Why, then, does this not happen? Why are unreasonable and nonbeneficial water uses not ended, if those engaging in them are acting without authority? Some would point to politics as the answer, but big-scale irrigators compose a small fraction of 1 percent of the state's voters. Some would point to public support for them and for land-based ways of live. This interpretation seems more weighty, but would public support exist if people knew that irrigators had no rights to act as they do and that irrigation practices were forcing urban dwellers to spend billions on alternative water supplies?

At bottom, irrigation continues for a different cultural reason. It continues because water rights are wrapped up in the language and aura of private property and citizens are inclined, or feel obligated, to respect private property. To respect valid rights is sound enough, indeed typically essential. But what needs respecting are the precise legal entitlements that property owners possess, not some exaggerated, cultural if not mythological characterization of them. Irrigators do not own water flows and never have. What they own are limited rights to use water flows so long as their uses are beneficial and reasonable and only when and so long as they are reasonable and beneficial. In *cultural* terms, owners have full control of what they own and can use their belongings as they see fit, without regard for effects on the common good. At *law*, in contrast, their rights are much more limited. But culture often trumps law, and it is doing so again in the arid lands of the American West. It is doing so, not the least because advocates for sensible water use have simply failed to engage the property-rights issue. They have sought ways to beg or cajole or economically induce irrigators to act more responsibly. Political power may well be at work, but the mythology of property, and the failure of reformers to challenge it directly, is much more to blame.

The life cycle of low-valued, desert irrigation practices has come to an end from any sensible economic, social, moral, and ecological perspective. If and when prevailing culture shifts, people will recognize this reality and perhaps wonder why they did not see it, and act on it, much sooner. In a similar manner, residents in the central United States may wonder why they allowed farmland owners for so long to overload natural systems with nitrogen and phosphorus, yielding dead zones in estuaries and a wide range of other ecological ills. They may wonder, too, why they allowed landowners to strip away vegetation right up to river banks and let mining companies chop off tops of mountains, pushing the "spoilage" into streams, to extract coal that ends up as atmospheric carbon and hazardous mounds of ash. We can blame the land degraders, but we need also to blame ourselves for not keeping property law in line with evolving values and needs.

Better Governance Systems

These various comments, about how property law might change, are usefully considered along with issues of institutional competence. To have a good system of private property over the long term we need good arrangements to calibrate and recalibrate private rights over time and to resolve the inevitable disputes. These arrangements are every bit as essential to a good property system as are the basic definitions of private rights. In general, multiple levels of government need to be involved in this work, operating at different spatial scales. The subsidiarity principle might properly be applied—power should be exercised at the lowest level of government that is able to use it effectively. The higher levels of government can set broad policy and prescribe the outline of private rights. Lower levels can add details. The specific processes used are critical. They need to be ones that get governing officials to think about larger contexts and the needs of landscapes as a whole. Officials also should consider the ways property works and how it can benefit us. One of the biggest dangers, institutionally, is what courts in the United States term "spot zoning," that is, decision-making that degenerates into parcel-by-parcel rulings that pay far too little attention to larger issues and contexts. Ideally, major policy decisions ought to occcur at fairly high levels, uninfluenced by the facts of specific local disputes.

The United States at the moment has an institutional mess on its hands when it comes to land-use decision-making. States have typically been the source of law that creates and defines private rights. It is the home of the common law of property. Local governments, though, have typically regulated land uses, with only vague guidance from the above. At the federal level, the Supreme Court decides constitutional issues—including regulatory takings cases—but has essentially no involvement in crafting property law and, in truth, does not show much understanding of private property as an institution or of sound environmental policy.[16] The levels of government do not work together. There is little or no coordination. It is no wonder that many people call for the creation of yet more government bodies, to do what existing governments fail to do.

The United States can offer examples of coordinated government arrangements, though not many. In the ecologically sensitive Pinelands area of New Jersey, for instance, land-use control is rather thoughtfully orchestrated by a scheme that involves the federal government, the state, and local governments, with the state setting broad policy and reviewing local plans for consistency with this policy. A few states—Oregon and Nevada, for instance—similarly have policy set at the state level and push local governments to develop local plans in harmony with them. But these are the exceptions and do not work as well as they might. As for local regulation, it is to varying degrees reviewed by state supreme courts. What courts mostly want to see, as they look over the shoulders of local regulators, is that land-use decisions are made in accordance with some sort of comprehensive land-use plan and that they meet certain requirements of consistency and uniformity. Courts push local regulators to look broadly at larger spatial scales and consider the public interest. They need to go further than they do, however, and push local regulators to think more directly about private property as an institution and about how their decisions affect its ability to promote the common good. Most of all, in the United States, we need to get states more involved. In the past, they have mostly just delegated power down, with few limits or duties. They need to set state land-use goals, and require that local plans be consistent with them. Overall, it is essential that states give more thought to the common good in the context of land use. They can then translate that understanding into the broad, recalibrated outlines of landowner rights.

When to Pay, and the Shifting Right to Develop

A rising issue in the United States (and elsewhere, it seems) has to do with the propriety of paying landowners to change or avoid land uses that are ecologically harmful or otherwise communally undesirable. When should landowners get paid to adopt new practices or forgo land alteration and when should the law directly prohibit unwanted practices?[17] In environmental terms, would greater or lesser payments promote better land uses?

A common line of reasoning among landowners and developers, anxious to get paid, is that good land use is beneficial to the community at large. That being the case, it is only appropriate for the beneficiaries—that is, taxpayers—to pay for these benefits. This reasoning has some merit, but it overlooks the critical first step, which is to define private property rights and, in doing so, ensure that the law allows only those private rights consistent with the common good. Property law, as just noted, has long prohibited land uses that are harmful, and appropriately so given that the legitimacy of property is linked to its ability to foster the common good.

Before getting to the issue of compensation it is appropriate and necessary, then, to recalibrate property rights to keep them up to date and in line with current social needs. Only after that is done—only after, for instance, redefining land-use harms—can we take the next step and consider compensation. It is not appropriate to pay landowners to forgo actions that are harmful. It is not appropriate to pay them to do their fair share in redressing carrying capacity harms or to avoid activities for which their lands are not well suited ecologically. On the other hand, it might well make sense to pay when a landowner is called upon to go beyond these baselines and to shoulder a burden not being imposed on other, similarly situated landowners.

The short answer, then, to the question of paying landowners is this: We ought to pay compensation when doing so would benefit the community as such. And the community would benefit when payment enhances the good effects of a well-calibrated system of private property. Payment, in other words, should enhance the public benefits of private property, of the types already mentioned. When probing this issue it becomes clear pretty fast that it is highly important for property law to protect *existing* land uses. We cannot expect the farmer to plant crops in the spring if she cannot harvest them in the fall. It is far less important that we protect hopes of developing land, or changing land uses, in the future. To get the farmer to plant crops, we do not need to promise a right to build condominiums on the land fifteen years from now. Accordingly, we should not halt *existing* land uses unless they are clearly harmful, some sort of public nuisance, or unless we compensate. Restricting the right to develop, on the other hand, is less problematic, particularly if we give advance notice.

When thinking about the right to develop, we need to bring in the issue of land values and how they arise. Imagine a piece of bare land, privately owned for years, which goes up in value because a city is built all around it. The land is now valuable, but where did the value come from? Not from the labors of the landowner. It came from the city, from the surrounding community. So if the community created the value, why does not the community get to keep it? Why is not this economic value a community asset?

Lawmakers in many parts of the world have put increasing restrictions on the right of private landowners to develop their lands. At some point one needs to ask: should we just do away with this right entirely as an element of private ownership? This would not mean doing away with development. It would mean getting rid of any presumed right for landowners to initiate new land uses. Then, as development is allowed when it makes communal sense, the owner would be expected to compensate the community for the rise in land value. This system, in fact, could work fairly well, and need not be a drag on economic growth. The effect would be to empower communities to decide where growth would work best. Landowners would have less say in the matter. Developers, though, might end up better off, as might big business looking to expand. They would simply need to have community planners point them where to go, and the development process thereafter might go more smoothly. Certainly groups wanting to acquire land for public uses, and even people wanting to go into farming in appropriate ways, would find this system much better because they could buy undeveloped land without having to pay prices inflated by the possibilities of lucrative land development. Park districts could add parkland; land trusts could protect valuable wildlife habitat and river corridors; farm families that wanted to keep land in farming or ranching could do so without undue pressure to give way.

If we are not willing to take that step, getting rid of presumed rights to develop, then at least we need to redefine development rights to take nature into account, to redefine impermissible land-use harms, and otherwise to tailor the rights of ownership so as to expect landowners to use their lands in ways that are reasonably sound ecologically.

In Sum: A Reform Agenda

Private property is a human, social arrangement that has taken many forms. Our ancestors shaped it and reshaped it over time to meet their needs, and it is entirely appropriate for people today to do the same. Indeed, we need to do the same to keep property morally legitimate and to keep it generating the benefits that it is possible to generate. Private property is a community tool. But it needs to be crafted well, with careful thought, to serve the community successfully. At the moment, property norms—particularly in modern culture, more so really than in law—are slanted so as to make environmental reforms more politically difficult, more culturally problematic, and in many places now more expensive.

In recent decades a rather thick fog has descended over the whole issue of private property, making it harder to see how property works, why it is morally problematic, and more. The case of water rights in California is merely a handy illustration. Any inquiry into it is easily distracted by claims that property is first and foremost a protection of individual liberty, and that it is a source of power that stands apart from public power. The confusions, as explained, are many. The environmental reform effort needs to counter the confusions and help Americans generally embrace an alternative ownership vision, one that achieves property's principal goals but does so at vastly reduced costs in land misuse.

The environmental effort needs to be asking: How can we define private rights today so that the institution generates the most benefit for us, particularly the benefit of helping us live prosperously on land in ways that sustain the land's long-term ecological functioning? That is the ultimate question in terms of legal change. As for strategy, what are the best ways to foster lines of public thinking about private property so that necessary legal changes make good sense? A strategy for the reform of private property, as cultural ideal and in law, might rightly include the following elements:

· Given that private property embodies and gives power to particular elements of American culture, property reform will need to take place as part of a larger push for change within modern culture, change of the type promoted by Leopold, Berry, Orr, and Pope Francis.

· Reform particularly needs to emphasize and strengthen the necessary link between morally sound property rights and the common good, a link easier to see when and as citizens understand the integrated natural whole, as Pope Francis puts it, as our common home, or as the land community (Leopold) on which our flourishing ultimately depends.

· The common good, which a sound property-rights system would promote, should properly reflect a well-considered distinction between the legitimate use of nature and the abuse of it; it would support the healthy ecological functioning of landscapes, while leaving due room for wildness, even as it also pays attention to human needs and social justice.

· Much reform effort will likely focus on the longstanding norm that owners avoid land and resource uses that cause undue harm to neighbors and communities. It will involve redefinitions of harm that rule out activities inconsistent with healthy lands and flourishing human life, including carrying-capacity harms. At the same time, however, reform efforts need to avoid undue stress on the negative, or what practices that must end. Appealing visions of what good land and resource uses can mean for human and other life need to be front and center. Accentuate the gains. The work here will, in effect, tailor the rights of landowners to take into account the natural features of what is owned. It will set as a guiding norm the ideal that uses of all parts of nature respect the particular ways that the parts function in ecological systems and in sustaining all life, though recognizing that the rightness of particular property uses needs to be judged at various spatial scales and that many sound land-use practices will disrupt nature at small scales.

· A new vision of property rights must certainly revive and give power to the property rights held by people collectively, rights in water, wildlife, beaches, and navigable rivers.

· Particularly in need of change is the now-outdated assumption that owners of land always have the legal right to develop it in some way that makes money for the owner. On this issue, as on others, the guiding star needs to be the common good. When and to what extent would the recognition of development rights in particular lands promote that common good? The issue is complex, to be sure. Answers need to reflect vital norms of equality and fairness. They also need to structure decision-making so as to reduce improper favoritism and corruption. On this issue, perhaps more than any other, it is wise to get back to the basics of private property as an institution, asking: Why do we have it? What benefits can it generate? And what mix of landowner rights would best generate those benefits while avoiding the costs and pitfalls?

· Good land use measured at larger spatial scales will inevitably require the concentration of human activities in some places while reducing the human presence elsewhere. The power of eminent domain (or *expropriation*, as it is termed outside the United States) will be an essential tool in that process of reconfiguration, aimed at, for instance, restoring the health of rivers and their floodplains, creating and protecting wildlife corridors, and ending intensive activities in sensitive coastal areas or above vulnerable aquifers. Governments hold the power of eminent domain as an inherent power of sovereignty, and always have. Reform efforts need to emphasize the legitimacy of this collective power, challenging the now-influential view that the power is somehow illegitimate. Property owners have always held their titles subject to the power of the lawmaking community to recall the property upon payment of just compensation. To give up the power, or avoid exercising it, is to give individual property owners yet more power to disrupt community plans and needs. It makes little sense to argue that private property as a whole is somehow enhanced when landowners are overpaid to give up their lands, given that overpayment simply takes extra money (property) from some property owners (taxpayers) and gives it to other property owners. Taxpayers as property owners are disrespected when they fail to get their money's worth.

· Private property is a potentially valuable social tool. It has been and can remain highly useful if the elements of ownership are wisely crafted and if lawmakers and those influencing them keep a sure sight on land health and other elements of the common good. As the story of California water law illustrates, private property in culture and law is out of step with the times. Much of it is back in the nineteenth century when resources seemed abundant, technology was less potent and threatening, and social values and ecological understandings were rather different. Reform is overdue.

SIX

Wilderness and Culture

Since the late nineteenth century the American conservation effort has featured a call to protect the nation's wildest places. John Muir gave the goal prominence over a century ago in his writings on Alaska and the Sierra Nevada mountains. Wilderness for Muir was a place for recreation and spiritual retreat, home to nature's beauty at its most sublime. After World War I animal researcher Victor Shelford convinced the newly formed Ecological Society of America to inventory the nation's dwindling patches of wild and to lobby governments to protect them, chiefly for scientific study. Aldo Leopold, then a forest ranger, explained the multiple values of wilderness in the 1920s, including big-game habitat and watershed protection, and helped convince the US Forest Service to designate the nation's first true wilderness reserve in New Mexico. The following decade Leopold joined a handful of professional colleagues to found The Wilderness Society, dedicated solely to the preservation cause. Pressures to protect wilderness areas mounted in the 1950s, particularly in more accessible wild places such as the Boundary Waters of Northern Minnesota. Working in the shadow of founder John Muir, the growing Sierra Club gave the issue particular emphasis. It led a coalition of citizen groups that, in 1964, finally convinced Congress to approve a Wilderness Act.

The flurry of pro-wilderness activity that preceded the federal statute was outdone over the next quarter century by the opposition to wilderness protection, particularly as the major federal land-management agencies, responding to orders from Congress (with great reluctance, in the case of the Bureau of Land Management), identified millions of federal acres suitable for further wilderness protections. The alleged benefits of wilderness protection were decried as costly, overstated, and elitist. The definition of wilderness itself was called into question. Controversy also erupted within the generally pro-wilderness side.[1] Some complained that wilderness areas were too small and that their boundaries reflected political factors, not ecology. Others fretted that wildness as a land-use norm had become too influential in environmental thought and distorted conservation efforts in inhabited landscapes.[2] A more stark complaint was that wilderness had become a cultural creation, lacking in objective meaning and sensible only to well-off people who used wilderness as playground. A similar complaint, mostly arising among academics in the humanities, was that nature itself had become a social construct, which meant that a call to preserve nature was confusing if not senseless. Other sideline critics highlighted how no part of the earth was fully untouched by humankind, given climate change and wide-spread DDT. This sobering fact seemed to discredit the whole preservation cause, or so some critics thought. Most recently the call has gone out that unmanaged wild areas are positively wrong-headed, given their vulnerability to human actions nearby and climate change's transformative power. To these doubters a hands-off attitude toward land makes no sense in an era of rapid natural change.

The upshot of this is that the term *wilderness* has become one of the more mischievous, elusive, and conflict-laden words in the English language. For many people, wilderness still shines as a yearned-for garden and a spiritual retreat. Just so for others, it lingers as a foe to oppose and dominate. In one view, wilderness is an embodiment of nature's time-tested ecological wisdom, worthy of respect and study. In another, it is a warehouse of resources that only the well-fed and misanthropic would forswear. Do wilderness areas really exist or is wilderness merely a socially constructed ideal of modern times? Do wilderness reserves reflect a mistaken view that humans exist apart from nature or are they instead a prudent form of survival insurance, places where we might, post-collapse, begin the search for more enduring ways to live?

Wilderness, and wildness generally, are central in any full consideration of the human place in nature. Aldo Leopold stressed the issue even as his late conservation efforts centered on farms. Wendell Berry celebrates wildness as well and wants it close at hand, in and around his pastoral home, not reserved in big enclaves far away. David Orr's views have resembled Berry's as he too has labored to improve developed lands, resource cycles, and (in Orr's case) industrial processes. Wilderness as such hardly appears in *Laudato Si'*, but the pope's call to respect all species, interconnections, and natural processes would require significant wildlife ranges, largely left untouched.

This chapter takes up the issue of wilderness and the ways wild places and processes can help make landscapes better for people and other life forms.

It dwells on wilderness also because of the revealing ways we have come to think and talk about wilderness. Wilderness is embedded in modern culture, in the culture that has struggled so to make sense of the human place in nature. To dig into its cultural forms is to expose further lessons about this culture and about its defining, and confining, traits. It is to see, for instance, why thoughtful people get confused about the language of social construction. It is to see similarly why many have sometimes used wilderness, unwisely, as a handy ideal and measure for good land use everywhere. Necessarily, it turns out, this probing requires a quick look at the origins of shared moral values and at the high cost we pay when insisting that public debates be supported by objective, empirically grounded evidence. It also requires a look into science—what it is, what it is not, and why we too often ask science to answer questions that go beyond its rightful range. One of the limitations of modern culture is that we have trouble seeing plainly our rightful role as value-creating beings. In truth we have the right and the power to vest nature with greater moral value—the kind of value so passionately expressed in *Laudato Si'*—and to revise our shared moral thinking to reflect that broader value. And we need to do so if our communities of life, humans included, are to flourish over time.

Ultimately, the question today is about the possible land-use benefits of wilderness preservation and how to gain them. Many of these cultural obstacles, though, need to be addressed first, before introducing these on-the-ground values. The way to begin this cultural inquiry is to say more about the vital line between the legitimate use of nature and the abuse of it, a topic variously raised in earlier chapters. The drawing of this line is a normative undertaking, not simply a scientific one, which is to say work that is linked for good and ill to core elements of a culture. It is difficult to think clearly about wilderness, or to justify it and manage it, without a fair sense of where the line is best drawn and how it is drawn.

The Use and Abuse of Nature

Like all living creatures, we humans need to interact with the surrounding natural world in order to live. As we go about doing this, we inevitably change the natural order, just as other species do. It is thus not inherently wrong for us to change the physical world; we cannot do otherwise. What is wrong—what is imprudent or immoral—is to use nature in ways that seem foolish or bad. All of this is clear enough.

More so than other species, people have the capacity to make nature better from their perspective, to improve upon it (beavers also do rather well).

This is most plainly true when it comes to gaining food and protection from harsh elements. Nature itself, of course, constrains this kind of ameliorative work. Oftentimes we successfully relax these constraints. But in important ways the earth operates as it does, and it is our evolutionary charge and challenge to find ways of living that respect the planet's ways and means. For the most part, we need to dwell on Earth in ways that respect its integrity and functioning.

When we take time to consider the point we are likely to recognize that the kinds of physical conditions on the planet that we refer to as environmental problems are, at bottom, unwanted conditions that people have caused. Nature's dynamic processes alter the world around us, sometimes in ways we find favorable, sometimes not. Nature can indeed cause havoc with its storms, tidal waves, volcanoes, droughts, and so on. An environmental problem, though, is something else, something other than nature acting or changing on its own. An environmental problem is a physical change caused by human conduct that we evaluate, in some sense, as wrongful or misguided. It is a condition that stems not from our legitimate use of nature, but from our misuse of it. And the problem, when we get to the real root of it, is not the changed natural condition—note the polluted air or the eroded soil—so much as it is the human behavior that underlies it.

This definition of an environmental problem—as human activity that degrades nature—is a useful one even though, and in part because, it is so plainly incomplete. When we define an environmental problem in terms of human conduct we draw attention to the conduct that is involved, if not to the particular people engaged in it. The definition helps also by highlighting our essential need to distinguish between the legitimate use of nature and the abuse of it. Without such a line how can we know which of the changes we make to nature are acceptable or good, and which are not?

We have long encountered trouble drawing this essential line between legitimate use and abuse, and, indeed, trouble even in appreciating the need for the line and the complexity of generating it. Decades into the environmental era, we still lack a clear enough sense about all of the factors and elements pertinent to this line drawing. Instead, we throw around the vague term *sustainability*, which lacks much meaning even when bolstered by the adjective *ecological*.[3] Sustainability is fine enough, but what are we sustaining in a world where nature itself evolves and human numbers and needs change as well? If we were to dig beneath the surface of many disputes about alleged environmental ills we would often find that what is in dispute is not factual evidence, not the science, but the proper standard for evaluating a landscape.

We can expose this intellectual murkiness by imagining a drive around an expansive farm landscape in central Illinois. Are the people here making good use of this naturally fertile place? Are they engaged only in the legitimate use of nature, or have they in some respects crossed the line to misuse it? Simply by driving around we would not gather nearly enough factual data to answer this question; there would be too much to learn, and the ecological effects would be challenging if not impossible to trace. But we would stumble on this question for the additional reason that we do not have, ready in hand, a sound normative standard to use in evaluating these rural land-use practices. Of course we need food to eat, as farmland defenders point out. But we need much more than that. A fully developed vision of good land use would reflect many relevant desires, values, and hopes, in addition to our basic need for something to eat. A landscape wholly devoted to food only begins to address them all.

As should be clear, two people who look upon such a farm landscape could have widely divergent ideas about whether the landscape suffers from severe environmental problems, due not to disputes about what is going on but simply to widely different normative standards. More scientific evidence won't narrow a gap of that type.

Objectivity and the Murkiness of Values

A big reason why we have difficulty distinguishing legitimate use from abuse is what might be termed our collective cult of objectivity, rooted in the eighteenth-century Enlightenment and strengthened over the generations by the rise of positivist science and liberal individualism.

When it comes to matters of public business we are prone to turn to verifiable facts and reason. We want arguments supported by factual evidence and we publicly test claims based on their logic, or at least we purport to. What gets pushed aside when we stick to facts and logic are claims based largely on emotion or claims that simply reflect a person's subjective view: a perspective rooted in contestable values, aesthetic preferences, and other variable beliefs.

In the current view one can act on subjective perspectives when making personal decisions, especially as a market consumer. But when it comes to public policy, we expect advocates to rise above their idiosyncrasies and stick to matters they can argue rationally if not prove. Objective proof can come inductively, based on empirical data that support factual claims. Instead, it can come more deductively, through reasoning that starts with established facts and that makes use in logical ways of some combination of fundamental liberal principles (e.g., equality, liberty, and fair process) and agreed-upon national goals such as economic growth, national defense, and public safety.[4]

The use of such liberal principles and shared goals in public discourse seems to come naturally to us, with little awareness that the principles and goals are themselves not really objective, not in the sense that we can prove them using facts and logic. We take our guiding principles and goals as obvious givens, as truths somehow embedded in the order of things. We speak about human rights—inalienable rights, as Thomas Jefferson termed them—not really knowing where such rights might come from and why we, among all species, possess them.[5]

We little consider why certain moral principles are embraced collectively and viewed as legitimate and binding while others are viewed as subjective, personal choices, ones that people might embrace if they choose but cannot force on others.[6] Facts can be put to empirical tests and verified to reasonably high confidence levels within the limits of our senses, testing equipment, and mental processes. But what about a claim that humans have the right to free speech, or that killing (or kicking a dog maliciously) is morally wrong?

As for where our shared values originally come from—not just our rules of morality, as commonly termed, but all of our standards of good versus bad or better versus worse—they have complex origins or points of beginning within individuals and among peoples. The literature on the topic is vast and rapidly growing, featuring, in recent years, important contributions on the origins of moral judgment from evolutionary biology and psychology.[7] To some extent our normative leanings are influenced by genetic factors brought out by social conditions.[8] They are influenced also by emotions or sentiments. Indeed, emotion seems to be the root base of most of our moral judgments, as David Hume urged in the eighteenth century.[9] Yet, even at the earliest stages of emergence, our normative impulses are also influenced and given shape by other factors beyond simple emotion or intuition. They are influenced by the facts of our material existence—ecological and social—and are guided by our powers of reason. In general, emotions, empirical facts, reason, and genetic influences somehow, in ways not well understood, combine within and among us to give rise to senses of good and bad and right and wrong.[10] They give rise in this murky way to social orders and cultures that, once in place and operating, exert great power over the thinking, feeling, and valuing of the people subject to them.

The basic lesson here is that the values and moral ideals we embrace together are grounded in our experiences and embedded in culture. In complex ways, we generate these moral values. They arise over time by social convention, albeit conventions commonly influenced by ecological facts, history, and our evolutionary trajectory.[11] Because we have trouble seeing this clearly, trouble recognizing our value-creating powers, we don't step forward to deliberate about them as we might, not in a context framed by due recognition of what we are doing. If we did that, if we more overtly understood our powers to generate new moral principles, then we might be better able to reshape our cultural values in ways that duly respect the natural order and our dependence on it.

We might illustrate where things stand, in terms of their efforts to promote new moral understandings, by considering claims made by animal-welfare advocates that certain high-functioning animals ought to enjoy moral value of some type.[12] Such claims routinely encounter resistance. Some critics push the claims aside by viewing the issue as simply a matter of personal choice: an individual can choose to view animals this way, but it is wrong to force others to agree. The issue, in other words, is not a matter for public resolution. Another line of criticism responds to animal welfare by insisting that welfare advocates *prove* their positions, using empirical data and logic.[13] Implicit in this reasoning is that moral values arise in this way, based on facts and logic alone.

Animal defenders often respond to this second line of criticism by attempting to do just what critics demand. They gather facts about humans and other animals and by logical reasoning using accepted moral principles seek to prove, logically, that certain animals possess moral value. The common, pro-animal argument goes like this: (1) Human moral value necessarily rests on some specific physical trait that humans possess, such as the ability to feel pain or to act intentionally (a factual claim); that being so, (2) moral value should extend beyond humans to all species possessed of the specific trait, as a matter of logic and by application of the accepted norm of equality.[14] The details of such pro-animal claims vary in terms of the specific trait or capacity that gives rise to moral value, but the format is pretty much the same. Facts and logic are mixed with the accepted moral value of humans and with the principle of equality to forge an argument in favor of moral value for other creatures.

It is easy to understand why animal-welfare advocates employ this approach. It is a means to extend moral value to at least some animals without challenging contemporary liberal individualism. Moreover, the proposal to extend value to animals clearly resembles (and is intended to resemble) prior extensions of moral value to include all people. Long ago it was typically thought that only certain humans had moral value, usually people who belonged to some tribe, religion, race, nationality, or the like.[15] Today, we commonly think instead that these distinctions are unimportant and that all humans are worthy. We have extended moral value to all humans. The next logical step, say animal-welfare advocates, is to extend moral worth further to certain nonhuman animals.

The problem with this reasoning is with the initial factual assumption, that we humans have value because of some particular physical trait or capability that we possess (for instance, because we experience pain as Jeremy Bentham said two centuries ago,[16] or perhaps because of our consciousness or self-consciousness). If that were true, if our moral value did arise because we possessed this trait, then the animal-welfare line of reasoning would work perfectly well. But what evidence is there that we have ever embraced Bentham's claim or any similar claim linking morality to a specific human capacity? What the record seems to suggest is that our moral value arises more directly. We recognize moral value in humans simply because they are human and because, these days, the differences among humans seem morally insignificant. We have moral value simply because we say that we do. After all, we don't go about testing people to see whether they do or do not possess any particular capacity. We treat the comatose as morally worthy, even though they are not conscious and feel no pain. Newborn infants have moral value; indeed many say value extends to the human zygote at conception. These stances would make no sense if moral value rested, for instance, on a high level of mental functioning.

If we saw more clearly how morality arises out of social convention it would be clearer to us that calls to extend moral value beyond humans really don't need to begin with an elusive search for the special human element or trait that generates moral value. We could openly propose that society revise its understandings of moral value to recognize value in a few, many, or like Pope Francis, all nonhuman animals.

Such a claim would be neither a factual assertion that certain nonhuman animals *do* have moral value nor a claim that, as a matter of fact and logic, they should have such value. It would instead be a call for us collectively to embrace a new norm to stand alongside the others, vesting or reflecting moral value in other life forms and in the interconnections that sustain them.

This animal-welfare story is also revealing because it highlights the modern tendency to think of moral value as something that inhabits an individual creature as a discrete being and doesn't exist in more diffuse form. The animal-welfare claim is that moral value resides in particular animals, and the claim is a sensible one (affirmed, as we've seen, by Pope Francis). But wild animals are embedded in natural systems and dependent on them, more so than humans in some respects. There is thus something awkward or incomplete about assigning value to them as distinct creatures—particularly, one might say, given predator-prey relationships—in a way that discounts or ignores these connections. The reasoning though, aligns well with the ways we think about ourselves as distinct, autonomous individuals possessed of a right to liberty and other significant rights.[17] Our emphasis on individual autonomy, though, clashes with the reality of the natural order, just as it does in the case of other animals. As Darwin showed long ago, we emerged out of the same evolutionary processes that gave rise to other life forms. We are unique neither in origin nor in the physical materials that compose our bodies. Our acts of daily living take place within ecological systems, with which we continuously interact. Our bodies, in turn, are inhabited by countless thousands of microscopic species, some of them dangerous to us, far more of them benign, and no small number of them useful if not essential.[18] As philosopher J. Baird Callicott explains, we might best understand our individual bodies not as single beings but as ecosystems, given the variety of life forms that exist on and within them—in the digestive tract, above all— and given the intricate interdependence of the parts.[19]

Given this interdependence it is misleading—even false—to claim that we are autonomous beings without immediately adding that we are embedded in larger natural and social orders and largely defined in essential ways by our interconnections. To be sure, we can and do move about, even hopping from continent to continent, and many of our ecological links are to places far away rather than near. But interdependence nonetheless is the inescapable reality. An individual-focused morality becomes suspect once people are embedded in systems and understood in essential part as elements of the systems. When the physical line between the individual and the surrounding world is thus blurred it makes sense then for moral value to spread more diffusely, applying to interconnections and natural processes as well as individuals. When we see ourselves embedded in larger natural wholes, in something like the land community of Aldo Leopold, it makes sense, as Leopold proposed, for the wholes as such to take on moral status.

The Cultural Muddle of Wilderness

How, then, does wilderness fit into all of this? How is our thought and action about wilderness influenced by these cultural traits?

We can start with the meaning of wilderness—what it is and whether it really exists. If we could understand more clearly that intangibles such as words, ideas, and values have no timeless existence to them, if we could see that they are all contingent human social creations, then we could put to rest much of the confusion surrounding wilderness. We would no longer worry about what wilderness "really means." We would no longer argue about whether wilderness really exists or whether it is, instead, a human social creation. (We might rightly still debate whether and to what extent humans have changed particular landscapes.) Further, we might see, better than we do, that the reasons for preserving wilderness are ones that we can and should generate ourselves.

Wilderness is, for starters, simply a word. Like all words, it is a human creation and has arisen over time, with varied meanings, through social convention.[20] As a word it is artificial in that we humans could have come up with some other word or words and could define them differently. Taken simply as a word, as a sign, wilderness does indeed exist; it is an intangible thing. But it exists within and among speaking humans and is inextricably linked to other words or signs.

Much the same can be said about wilderness as an idea or an ideal, apart from the words used to label it. All ideas and ideals are human creations, putting to one side what goes on in the minds of other creatures, and are artificial in the sense that we could have generated other ideas and other ideals. As for whether wilderness exists as an actual, physical place the answer depends on our definition. We could define wilderness so that hundreds of millions of acres of land and water satisfy the definition. We could define the term instead in such a way that no place on earth fit the definition.

As for the values associated with wilderness, they are, for the reasons already noted, also among the categories of intangible things that come from people, whether we create them or profess merely to recognize them. Are wilderness areas worth protecting? At one level, the answer is simple: they are if and to the extent we say they are, though we should immediately add that a failure to protect wilderness could have effects that we (or our descendants) dislike. The concept of value means, in practice, valuable to a sentient creature that is capable of attributing value.[21] There is no value without someone to create it or at least recognize it.[22] The value- or norm-forming process, for the reasons covered, is a murky one that in some way draws upon facts and reason, as well as sentiments and genetic influences. The natural world thus plays critical, indirect roles in the value-creating process. Through careful study we can discover new facts about wilderness-type places. These facts, in turn, might influence our normative judgments about the places.

The human agency in forming of values is easier to see when we talk about *instrumental* values—about the ways wilderness areas and other components of nature are valuable because they promote our flourishing. We play a role in deciding how we want to live. Yet human origins also underlie what is termed *intrinsic* or inherent value. While intrinsic value exists (by definition) independently of any direct contribution to human welfare, it remains every bit as dependent as instrumental value on a human to embrace the value. Intrinsic value simply means value that humans recognize for reasons or sentiments that are not directly linked to the promotion of human welfare.[23] We can say that a gorilla has intrinsic value apart from any benefit we might get from it. But the gorilla nonetheless has such intrinsic value in our value scheme—the only scheme that we can know—because people have come to value it.

When thinking about wilderness in normative terms, we could begin with the question: How might we best define wilderness as a guiding land-use ideal? It is a fair enough phrasing and might promote good thought. But the word *wilderness* is not a necessary component of this question and it is diverting to frame the question in this way, as chiefly definitional. A better approach—one that puts the norm-setting work front and center—is to ask instead: How might we best make collective use of a particular landscape, the one we might call wilderness? How should we live in this particular place over the long term? Or perhaps better still: How, in this landscape, should we draw the line between legitimately using nature and abusing it?

The benefit of these latter questions is that they present the fundamental challenge in terms that are clearly normative, rather than definitional, rightly implying that answers are ones that require us to make choices. These questions are not answered by gathering and synthesizing data, nor can we resolve them using reason alone. Further, these latter questions usefully imply that our answers should be based on all-things-considered assessments. Of course we need food to eat, to return to our farmland example. But we misuse land when we devote all of it to food production, leaving no parts of nature to satisfy our other needs and values.

When we head down this path of identifying possible values for wilderness, what we are likely to produce are normative reasons to value and protect wilderness that roughly fall into two categories. We have reasons that relate directly to the wilderness areas themselves, as distinct places. And we have reasons to care about wilderness areas when considered as parts of larger landscapes and systems.

Exercising Our Powers to Define and Value

Just as modern culture can muddle our thinking about wilderness—muddle our efforts to get clear on what it means, whether it exists, and so on—it can also complicate this essential work of distinguishing the legitimate use of nature from abuse. A common tendency, when we do recognize the line-drawing task, is to view it as one for science to resolve. But science is not up to the task, and we need to know why. Science is essential to the work, but to put science in charge is to curtail severely our collective norm-setting powers.

Putting science in its place. Particularly on environmental issues we are prone to insist on objectivity and expect scientists to make sense of them. We turn to science, we insist that decisions be based on science, even when the fundamental issue is a normative one on which science as such has little to say. Science is a tool for finding and testing facts about the physical world and its operations. It can tell us, for instance, whether our actions are having the effect of changing the climate and, if so, how that will play out in physical terms. But it cannot tell us whether the resulting effects are normatively good or bad, not without importing and employing normative values from outside science.[24] In the same way science cannot generate the standard to use in evaluating whether Illinois farm landscapes are being well used, even though the facts it supplies are essential and it can point immediately to ways that industrial farming saps fertility. Science is simply not in the business of creating and applying normative standards, though individual scientists, taking off their science hats, might be quite good at it.

When scientists are on center stage, viewed as the experts, citizens get pushed aside. They never really engage the normative, line-drawing work in any sensible way. At that point it becomes easy to embrace a simplistic perspective of one form or another. One option is to insist that humans are conquerors of nature, and that individuals, especially businesses, can freely use nature as they see fit so long as they avoid causing measurable, direct harm to the person or property of other individuals. A much different option, equally simplistic, is to honor nature and its beauties and to presume that all human change is suspect if not abusive.[25] Both approaches, it should be clear, are deficient. The liberty-based perspective ignores numerous normative factors that should enter into any all-things-considered assessment, including claims that other life forms and future generations should carry moral weight. The latter, left-leaning approach is equally simplistic in that, by equating all or nearly all human-caused change with degradation, it views the human presence on Earth as inherently bad. It simplifies the line-drawing by labeling all human change as abusive.

Despite occasional claims to the contrary the environmental movement has rarely taken this second approach. It has rarely used unaltered nature as its benchmark of acceptable land use. Environmental statutes—the Wilderness Act excepted—give no special status to unaltered nature.[26] The Clean

Water Act does propose the lofty goal of reducing pollution discharges to zero, but neither its water quality–based approach to pollution nor its technology-based standards pay much attention to this nonbinding prefatory vision.[27] Most often, environmental laws and policies are expressly written to reduce harm to people and their activities, not to protect nature independently of people.[28]

Even so, this all-change-is-abusive perspective remains a cause of sloppy thinking. It is easy for defenders of nature, loving it as they often do, to slip into language that treats all change as bad or that measures the degree of human-caused harm to nature starting with unaltered nature as the baseline. In much the same way, it is equally easy for defenders of intensive land uses—defenders of industrial-style agriculture, for instance—to insist that lands are well used whenever they produce food that humans eat, as if a responsible normative evaluation could consider one factor alone.

The many possible definitions. When it comes to wilderness, the impulse of many is to assume that we should leave it untouched. We should not intervene in wilderness areas, even to slow or reverse changes within them caused by human activities elsewhere.[29] But we have good reason to move cautiously on this issue. This alluring, do-not-touch approach is easy to reach logically with a quite simple syllogism: (1) define wilderness as an area unaltered by humans; (2) designate a particular place as wilderness; and then (3) decide that this place should be left untouched. The logic is sound. But where along the way do we give thought to the full range of normative factors relevant in deciding how to live in particular places? Where do we stop to ask directly: Why are we valuing particular places that we are prone to call wilderness?

We could exercise our word-creating powers to define wilderness as an area completely unaffected by humans, and do so even though no such places exist on the planet. There is nothing especially wrong with a word that refers to an unachievable ideal. Geometers refer to the ideal of a circle even though in physical terms there is no perfect circle, given the vagaries of molecules and atoms. Similarly, we speak of justice though we have yet to achieve it and do not expect to. On the other side, wilderness could be defined— as it long was—in ways that allowed for minimal human alteration. Aldo Leopold, an early advocate, believed that minor intrusions and structures were consistent with wilderness preservation.[30] For Sigurd Olson, the lyrical wilderness advocate of the Northwoods, wilderness was more a matter of aesthetics and emotional appeal. He sensed that a primitive trapper's cabin added to the wilderness feel rather than the reverse, and that a few scattered resorts in the canoe country did not materially diminish wilderness values.[31]

To say that wilderness can be defined in varied ways is by no means to say that individuals as such should be free to pick their own definitions. To allow individuals to define the term separately is to vest them with the power of making the foundational normative choices. It is to give in to the cultural ideal of liberal autonomy. It is to treat wilderness values not as shared normative axioms but as mere subjective preferences. To the contrary, the normative work that needs doing is collective work, with the results incorporated into shared language, thought, and policy.

Wilderness and Good Land Use

In what ways, then, might we collectively attribute value to wilderness areas as we exercise our powers as value creators? What values might we assign to such areas considered as isolated places? Similarly, how might we value them when considered as components parts of larger landscapes, which also include places where people live and work?

Valuing wilderness areas in isolation. We have many reasons for deciding, collectively, to invest little-altered lands with intrinsic value, respecting them for what they are, apart from any identifiable benefits they provide to us today. We could value them so that future generations of humans might enjoy and benefit from them. We could recognize that wilderness-type places could well provide benefits in the future that we cannot now see and thus cannot value directly. Our moral thinking could incorporate senses and visions of virtue that we implement by showing restraint when altering nature. Wildlands protection could arise out of virtues that are best implemented by treating wild places as intrinsically valuable. More directly, we could respect the wild creatures that live in such places, and the unique biotic communities that they help compose, and treat them with moral reverence.

These reasons for recognizing intrinsic value are largely consistent with instrumental values for wild places that involve modest human uses of them. Wildlands provide special recreational opportunities. For many people they provide places for spiritual retreat. For scientists they provide places for ecological study; indeed, the first major push to protect wild places, as noted, was by Victor Shelford and other ecologists who sought to protect them for future study.[32]

These limited, gentle uses of wild places—when added to the reasons for attributing intrinsic value—push for a definition of wilderness as a place largely unaltered by people or at least where future alterations are strictly limited. But these various factors have flexibility to them, and they hardly give reason to withhold protection from places that have been noticeably altered. We can benefit future generations—to return to the first rationale of intrinsic value—by preserving the best examples of biotic communities even when they do show noticeable human change. We can similarly protect prospects for discovering new instrumental uses of wild places and their biotic and abiotic components—another protection rationale—in places that display distinct human presence. Indeed, the entire list of preservation rationales given so far would seem to apply to landscapes that humans have certainly altered, particularly when such landscapes are the best we have available. That being so, we have ample reason to define wilderness so that it includes such valuable lands, despite their human change, and to set management limits that reflect the reasons why we are protecting such lands.

The landscape-scale benefits. Beyond these bases for intrinsic value, and these limited direct human uses, we have the various other instrumental reasons for protecting wilderness—those having to do with the ways wild lands promote the good use of surrounding lands that people inhabit and alter more significantly.

Scattered patches of wilderness can help promote ecological health and functioning in the larger landscapes of which they are parts. They can provide room for plant and animal species, enriching biodiversity and enhancing the functioning of larger landscapes. They can also help sustain the healthy functioning of hydrologic systems by maintaining good water flows and water quality, thereby protecting and building fertile soil and aiding a wide suite of life forms. More generally, wilderness-type enclaves can help the humans who inhabit larger landscapes to stay on the right side of the use-abuse line by promoting the good functioning and biological composition of the larger landscapes.

A normative standard at the landscape scale. For wilderness to be understood this way the normative work described above needs to be done. A sensible line needs to be drawn between the legitimate use of a landscape and the abuse of it, taking into account all relevant factors. Without a reasonably clear line it is hard to know how wilderness preservation might help promote good land use at the larger scale. *Within* a wilderness area, as noted, the guiding land-use standard might well be one that tolerates only very modest uses and changes. But such a standard—perhaps based on some definition of ecological integrity—is not well suited to apply in the larger landscape of which a wilderness area forms a part. We need to use many lands much more intensively than this, to live, grow food, and much more. For such occupied lands, we need a much different guiding vision of good land use, one that allows for intensive change while also ensuring that the land's basic fertility and productivity are kept intact. This landscape-scale vision, going further, should take into account nature's own dynamism, our limited knowledge of nature's functioning, and the prudence of acting cautiously so that we avoid costly, irreparable surprises.[33]

The bottom line is thus that, when we think about wilderness areas (and, by similar reasoning, wildlife refuges), it is sensible to come up with two normative land-use standards, one that guides actions in the wild places themselves, the other that guides uses of the larger landscapes that includes places where people live and work. The internal standard for wild reserves has long been the subject of much thought and writing, and legal guidance reflecting it is set forth in the Wilderness Act and, in the case of national wildlife refuges, in the National Wildlife Refuge System Improvement Act of 1998.[34] Preservation advocates have pressed hard for statutory standards that minimize human alterations of these places, with much success.[35]

Far less attention has been paid to the second standard of broader application, although much good writing has been done on ecological functioning and on specific limits for pollutants.[36] This deficiency is critical for wilderness protection because the landscape-scale standard is needed in order to measure the full benefits that wilderness areas can generate. How do particular wilderness areas help the residents of landscape-scale landscapes use their lands rightly? Similarly, what largely wild lands might best be protected to achieve these landscape-scale benefits, and how could the wild lands generate them fully? Clear thinking here could assist in both identifying lands to protect and setting standards for management and use. It could also help wilderness defenders and managers to explain much better the full range of instrumental values that wilderness areas provide. Defenders could point not just to the ways that wilderness areas are directly valuable to humans—as they already do—but also to the ways that wildlands help promote the more encompassing goal of good land use at a larger spatial scale.[37] Who knows, these larger-scale benefits arising out of ecological interdependence and interconnection might well be more valuable than the benefits that come from using wilderness areas directly.

The Work of Protecting Wild Places

The conclusions reached above come together readily to form an agenda for moving forward on wilderness protection. The tasks, in something like this order, are as follows:

· The first step is to cut through the definitional fog and explain that any definition of wilderness is based on social choice and that we are wise to define the term in whatever way proves most helpful. It is pointless to debate what wilderness might really mean and whether it really exists. It is pointless also to debate whether wilderness (or nature) is somehow a social construct: Wilderness as word and idea certainly is; wilderness as place is not. Having claimed this power to define wilderness, though, the actual work of defining it is best delayed a bit, until the benefits of wilderness are more clear and a definition can be calibrated to promote them.

· The second step is get to work on the task of formulating values and standards to distinguish decent land use from abuse (work that includes, as noted, putting science in its rightful place and getting clear on the origins of morality).

· At this stage, with a sense of good land use overall, it becomes possible to think clearly about the many ways that wilderness areas might help promote good land use, if not be essential to it. Land uses are best evaluated at varied spatial scales, including very large ones. Only at those larger scales can one judge the full benefits and costs of wilderness reserves. The larger scales are also appropriate when selecting lands for protection and identifying how protected lands, if rightly managed, might best generate larger-scale benefits.

· The process of spotting benefits from wilderness preservation, considering the large scale, can then bring in the benefits that come from the wilderness areas as distinct enclaves, including benefits for visitors, scientists, and the many life forms that need wild places to survive. Once all the benefits become clear, it should be easier to craft management standards for the reserves, keyed to their benefits. It may make sense in some settings to take something like a complete hands-off management policy. But there is little reason to privilege that management approach, little reason to assume that it will be better than alternative ones.

· While it may seem odd, the work of defining wilderness, in terms of setting the criteria for selection (the main legal function of the definition), might wisely come as the last step. At that point we might have a fair grasp of the possible benefits of protecting undeveloped lands. It might be true, at that point, that we want to protect lands that people have altered to some degree and thus might want a definition of wilderness that includes them. Such lands might do just as well as untouched lands in sustaining ecological functions and biodiversity and in providing needed recreational opportunities. One option is to propose gradations of wild lands, akin to the gradations of protected rivers under the Wild and Scenic Rivers Act.

Wilderness and Culture

Particularly in Congress the reform effort to improve our uses of nature has largely run aground. More localized labors, and federal administrative initiatives, continue to move ahead, but on key problems—climate change, water overuses, spreading dead zones, biological declines—the pace of reform is much too slow. Delays and resistance are not due to lack of good science and technology, although better science and technology can always help. The resistance lies deeper. It lies in modern culture, as Leopold, Berry, Orr, and Francis have all insisted, in the ways we see and value the world and think about our place in it. It lies in our confusion about values and our rush to positivism. Given the need for cultural change, the question arises: Might wilderness areas and preservation work help bring it about? Could wilderness areas be instrumentally valuable in this important respect as well, as tools to broaden our moral sensibilities and embed us better in nature?

In the search for answers we might begin with Aldo Leopold's final writing on wilderness, the essay, simply entitled "Wilderness," included in his posthumous collection, *A Sand County Almanac and Sketches Here and There*.[38] Readers of that famous book know that it ends with the much-quoted essay "The Land Ethic," which, given its placement and breadth, is typically viewed as the culmination of his conservation message.[39] But it was only after Leopold died, during the editorial process, that "The Land Ethic" was moved to the end of the book.[40] In Leopold's manuscript, "Wilderness" brought the book to a close, and a rather different one.[41]

Leopold's "The Land Ethic" wound down with a quiet, reflective passage. Our land-development technology had many good uses, Leopold observed in the final paragraph, and we were not about to give it up entirely. But we were in need "of gentler and more objective criteria for its successful use." Had "Wilderness" remained where Leopold placed it, the book would sounded a more somber final tone. In the essay's concluding sentences Leopold lamented the sad plight of "the shallow-minded modern" who had "lost his rootage in the land." Big-scale human history was a saga of overlapping, failed efforts to find ways of living in nature that sustained the land's fertility. When failure came, as it had in many civilizations, people had to regroup and set out again to resettle the land, organizing "yet another search for a durable scale of values." "Raw wilderness," Leopold urged, could supply "definition and meaning" to that search.[42] To protect wilderness was thus to protect the option for second chances, to protect the chance to attempt again to craft ways of living consistent with the cycles and means of nature that wilderness areas illustrated. By durable values Leopold surely meant better ways of understanding our ecological plight in the natural order. He meant valuing the elements and processes of nature that kept the land fertile and productive, thereby sustaining civilization.

As wilderness advocate Howard Zahniser put it a half-century ago, we need to sense that we are "dependent members of an interdependent community of living creatures," members whose knowledge of the world is interspersed with a great deal of ignorance and whose deepest yearnings—our inbred normative hopes—are to live in ways that sustain life for generations.[43] Few places exhibit interconnection and interdependence more vividly than wilderness areas. Few actions show more humility and a greater admission of ignorance than the work of setting wild places aside. Few acts proclaim a new normative value more vividly than a shared effort to respect and defend what we once treated as an object of conquest.

In the end, this cultural-reform component might well be the most important connection between wilderness and culture. Wilderness is not really needed to learn basic scientific lessons of ecological functioning; those we know, and we can learn more about them elsewhere. Our greater need is to stimulate our moral imaginations. It is to reframe our understandings of ourselves as living creatures. Interconnection is the guiding norm, not autonomy. In our interdependence we are on an even plane with other creatures, not some special life form. The earth's crust and lower atmosphere form a community of life of which we are members, a community that can be more or less healthy in its functioning and fertility. These are the basic facts and they need to shape our cognition, consciousness, and moral choices.

We can use wild places of many types—more than just those that now qualify for legal protection—to illustrate and help instill these vital cultural understandings and values. We can use them, in sum, as mechanisms to help stimulate and guide cultural reform. But to do that much else must also be done, and done much better than we have yet done it. Once we see the kind of cultural change that our present plight demands, wilderness and its protection could well help us achieve it.

SEVEN

Naming the Tragedy

Since its appearance in 1968, Garrett Hardin's short article in *Science*, "The Tragedy of the Commons," has become an especially handy source for scholars to cite in support of an array of claims about nature and why we misuse it.[1] In his article, Hardin drew attention to the rising human population and offered an explanation why, absent intervention, it would keep rising, well beyond the planet's carrying capacity and even when it brought suffering and degradation. To illustrate his explanatory theory of excess Hardin included a tale about a grazing pasture that suffered tragic decline because of overuse. It was this short, fictional narrative that drew great interest and soon turned Hardin's article into a classic.

In Hardin's story, individual cattle grazers were free to use the pasture as they liked. They could add more livestock at any time, and did so even when the extra animals caused overgrazing and degradation. An individual grazer had an incentive to act this way, to add an extra head, because the forage eaten by the animal benefited the grazer. The additional animal brought net harm due to the overgrazing, particularly as other grazers followed suit. But that harm was spread among all grazers while the benefits of the extra animal went to the owner alone. Each grazer thus had an incentive to act in ways that brought tragic consequences to the landscape and its users. For the "rational" grazer, Hardin contended, adding more animals was "the only sensible course." And it was individual freedom that made it all possible. As Hardin famously contended, "Ruin is the destination toward which all men rush, each pursuing his own best interest in a society that believes in the freedom of the commons."[2]

Hardin described his grazing region as a commons, and it was, in the sense that many people shared its use. More precisely the region was an open-access commons in that no norms or rules limited the ability of any grazer to graze more animals at will.[3] The outcome of this freedom, Hardin asserted, was tragic in that it led inexorably to misuse of the pasture and harm to the grazers themselves. Hardin did not pause to define good pasture use; he did not explain how he would distinguish between the legitimate use of a grazing region and the misuse of it. His was a simple tale, with no need to get specific. At some point, overgrazing reduced the region's forage productivity, an outcome he deemed bad.

Hardin's conclusion was that this kind of selfish freedom needed to disappear. In some way lawmakers needed to limit it through coercive means. As a democrat, Hardin believed that binding limits should be "mutually agreed upon by the majority of the people affected," not imposed by autocrats.[4] He thus phrased his solution as mutual coercion, mutually agreed upon. Coercion could take the form of something like governmental regulation. It could also come, in the case of the grazing tragedy, through the division of the pasture into privately owned shares.[5] If the latter was done, the ill effects of overgrazing by any individual grazer would be felt by the grazer alone, not shared by others, thus aligning the costs and benefits of overgrazing and leading, presumably, to less or no overuse. Hardin presented these remedial options as variations on the theme of mutual coercion, but many readers would treat them as more distinct—a public ownership-regulatory option and a private-property option.[6]

Over the years Hardin's tale has become something of a Rorschach test, akin to the personality test developed by Swiss psychologist Hermann Rorschach in which patients are shown inkblots and asked to describe what they see. In much the same way, readers of Hardin's tale can come away with widely varied interpretations. What seems important in this story? What truths are displayed? And what omissions or errors might be embedded in it and in Hardin's explanation?

The possible answers to these questions are many, and very likely turn mostly on traits that a reader brings to the provocative narrative. Based on answers to such questions one might well discern the political leanings of a reader and probably more: his ecological understandings, his thoughts about government pro and con, the value he places in standard neoclassical economics, and his views of private ownership and its benefits. Indeed, one might guess that answers from readers could be spread out along a graduated spectrum. Does the tale sum up our environmental predicament? Is it proof of the value of private property and support for more of it? Is it proof of the ill effects of allowing people to act free of control? And might it contain still other lessons about human nature, economics, and our planetary plight?

Hardin's tale is so malleable in part because Hardin presented it simply and most readers have reacted with equal simplicity. The story, though, can also be teased apart more carefully, with particular regard for embedded assumptions and for issues that, though raised implicitly, were not flagged or probed by Hardin himself. The story, to be sure, has to do with ecological decline. But it takes digging to get to the bottom of this story, to figure out both why the degradation takes place and what steps the grazers (or others) would need to take to avoid the decline. On both points—the root causes of degradation and the steps needed to avert it—Hardin's story is radically incomplete. To fill it out is to gain considerable insight both on the causes of today's environmental ills and on the reasons why modern society has such trouble coming to grips with them.

Three Basics

Readers of Hardin's tale—of all political and cultural stripes—ought to find agreement on three basic claims about the tragedy. It is useful to begin with these claims before moving on to points that are less clear.

For starters, the degradation of Hardin's grazing region was caused by the grazers themselves, by human conduct. The cattle, to be sure, ate the plants and fouled the waterways. But the grazers introduced the animals and controlled them. People misused the region, not bovines. It is an essential point and foundational. We gain clarity by pointing the finger at the people causing the harm, as Hardin did, not at the harm itself.

Second, few readers are likely to think grazing is always ecologically bad, even if they object to treating cattle this way. That is, some level of grazing is acceptable in some places.

Certain grazing entails the legitimate use of nature by people, while grazing at a higher level or in the wrong place crosses the line and becomes abusive. To figure out whether grazing is excessive thus requires a line between legitimate use and abuse. All grazing brings changes to a landscape so this line-drawing in effect distinguishes between changes made to nature that are acceptable or ameliorative and changes that instead are unwise, immoral, or otherwise misdirected. As considered in the last chapter, this line-drawing is very much a normative task, even as it makes extensive use of scientific facts. It is up to people to decide, based on their values and normative preferences, where the line should lie. Nature by itself does not draw it; it merely reacts to what humans do, particularly when we ignore natural limits. Nor can science alone draw it even as science might tell us the consequences of alternative grazing options.

The third basic lesson embedded in Hardin's tale is that the grazers collectively would be better off if they got together and came up with coercive rules limiting their individual uses of the commons. Pretty much all readers can see that the grazers ought to do this, and that they likely would do it given adequate opportunity. By implication, the "rational" decision that a grazer might sensibly make acting as an individual—to add more cattle— would likely differ from the equally rational decision the same individual would make when joining with other grazers to set up a governance regime. Working with others the individual could vote to impose limits that would keep him from doing exactly what he would choose to do as an autonomous individual.

We need to put the last point plainly: the decision a grazer makes as an individual autonomous market actor could differ radically from the decision he would likely make in his role as citizen-lawmaker. Both decisions reflect what the grazer wants. Both also make good sense in simple economic terms; they are, in that regard, intellectually coherent. In short, the preferences of grazers—the preferences of people generally—can vary based on the role they play. And they might vary because, when the grazers cooperate, they have options available to them that are not available when acting as competitive individuals. Further, social decision-making often requires individuals to justify their desired actions in a public setting, opening them to challenge, criticism, and possible amendment.

This divergence has long been known as the citizen-consumer dichotomy.[7] It ought to be, but for the most part is not, central to nearly all talk about our environmental plight. Similarly, Hardin's tale ought to stand, though it rarely does, as the poster case refuting the common claim that people show their true wants when they spend their own money as individuals, not when they tell pollsters how they would vote.

What Is a Commons?

One hornet's nest that Hardin stirred up arose from his use of the term *commons*. In the case of population growth, he meant the term to refer to the entire planet, which was a commons from the perspective of people because they all shared it, albeit in highly unequal ways.[8] That use of the term drew little objection. What became contentious was his use of the term *commons* to describe the grazing region. The pasture was also a commons, but it was one that scholars would term an open-access commons, one that grazers could use free of any norms or limits.[9] There are such commons in the world—the atmosphere for the most part, and many fisheries.[10] But long-time grazing commons are typically places where users are embedded in collective governance regimes with rights prescribed by norms or other rules. Such a place is also termed a commons. As scholars pointed out, a well-managed commons of this type would produce not tragedy but something more like its opposite.[11] It could yield maximum benefits for the grazers, better even than if the commons were divided into separate geographic shares, with nothing like the decline that Hardin predicted.

This terminology objection, of course, did not really challenge anything Hardin had said. Indeed, he had expressly noted that users could avoid the tragedy by embracing a coercive management system.[12] The problem, then, was not with Hardin but with simplistic interpretations of this tale, particularly by readers who liked the idea of fragmenting nature into privately owned shares. A commons led to tragedy while privatization led to lasting productivity: that was the simplistic interpretation, one that students of real-life, successful grazing arrangements found wrongheaded.[13]

A more general concern about Hardin's use of the term looked to his implicit claim that a commons was in some way a special kind of place. It was a place that had not been divided into private shares. When a commons was divided, when each part of it had a distinct owner, then it was no longer a commons. Hardin, we might note, did not say this exactly. Indeed, his population example pointed in a rather different direction, to the view that the entire earth remained a commons despite fragmentation into nations and private shares.[14] But again, Hardin's story was useful to many types of readers and it was easy to distinguish, using his grazing tale, between landscapes that were divided into private shares and those that were not. Only the latter were subject to tragic misuse.

One objection to this interpretation (that is, to the claim that privatization solves the tragedy) is that it overlooks ecological interconnections. What happens on one parcel is linked to physical conditions elsewhere given flows of air, water, wildlife, and nutrients. One landowner is enough to break up a wildlife migration corridor. One landowner can alter drainage in ways that substantially disturb downward owners. To the extent of interconnections, unseen ones as well as the seen, a landscape remains a commons even after its division.

A related objection is that landowners everywhere participate routinely in an economic commons when they produce goods or services for the market, competing for customers who move about as freely as cattle.[15] Here we can consider the typical city in which owners of gas stations or drug stores compete for customers, establishing businesses in new locations in a never-ending quest for market share. The competition is plainly wasteful; there is no need for similar gas stations or drugstores adjacent to one another. And waste means ecological cost—tragedy—in some place, somewhere. Further, people living in the area are forced to drive past the competing businesses, and countless ones like them, traveling further distances simply to get where they want. City needs could be adequately met with fewer gas stations and drugstores, particularly abandoned ones. Too many of them populate the urban area, which remains a commons, despite its fragmented land ownership, due to the flows of goods, services, workers, and customers.

The same lesson about overuse is illustrated in the cases of oil fields and groundwater aquifers. Oil producers can often exploit an entire field using only one or a few wells.[16] Water pumpers can access an aquifer with similar efficiency.[17] When the overlying land, though, is owned by large numbers of people and each has the right to install a well, the number of wells can quickly become excessive and wasteful. In the case of oil wells, too-rapid pumping by too many wells can reduce overall output by releasing pressure too quickly. If we view the underlying oil deposit or aquifer as the common asset, one can argue that it has not been divided into private shares simply because the overlying land is fragmented. This is true enough, but what it illustrates is that division of lands into private shares almost never brings the commons to a full end. So long as private parcels are not completely sealed off, with no spillover effects among them—including no movement of people or goods among them—then features of a commons remain. A city remains a commons so long as people and their things freely move among competing locations, much as oil and water migrate among wells.

Hardin's tale, in short, is not usefully read as a story about a special kind of place known as a commons, a place distinguishable from other places that are privately owned. Pretty much all lands and resources face the dangers of tragic misuse whenever and to the extent activities unfolding on them trigger effects elsewhere. As for sealing off each parcel so that spillover effects end, the idea pushed to its fullness is nonsense. People need to come and go, as do the goods they produce and need. The health of lands everywhere depends on ecological systems and processes that necessarily transcend boundaries. To sever or disrupt them is to invite tragedy, not avert it. Put simply, a commons exists in any setting characterized by interconnection and interdependence, whether ecological or social, which is to say essentially everywhere.

Privatization as a Solution

Hardin's discussion of the solutions to his tragedy is subject to more direct challenge because of his too-easy assumption that the simple division of the grazing commons would end misuse. Hardin did not dwell on the point; he offered the observation mostly as an aside. But it was a claim that many readers found congenial, particularly readers who disliked government and viewed the market as a superior mechanism for resource-use decisions.[18]

The weakness of this stance is easy enough to see simply by looking to landscapes in private hands. Overgrazing afflicts private lands as well as public ones. Private farmlands have long suffered from soil erosion and degradation. In the American Midwest, farmers on private lands routinely spread fertilizers and pesticides, leading not just to topsoil degradation and declines in soil biological diversity but to massive problems in waterways (dead zones most vividly). Too often owners of private forests have clear-cut them in ways that cause soil declines, siltation, fish kills, and other ill effects.

An individual owner could, to be sure, take good care of land, and many do, putting to one side land-use goals that require large-scale coordination. But poor land uses are common, and it is no answer to point out that many costs of degradation are borne by the owner.

There are, in fact, a wide range of reasons why owners of private lands do not and sometimes cannot take good care of them.[19] The owners could be unaware of the effects of what they are doing, given that many harms are invisible, distant, or slow-emerging. They could act based on strong competitive pressures or simply know no other way to behave. An overall culture of land misuse—a culture of refusing to recognize nature's limits—can exude an aura of legitimacy, carried forward by tradition and aided by misinformation dispensed retail by (for instance) equipment and materiel suppliers. Rational actors could use a discount rate for future costs and benefits that leads to exploitation today with the proceeds then invested elsewhere in higher-yielding assets. A short-time horizon might also be used for other reasons—simply the advanced age of the land user, for instance. The reasons are many and have long been known.

Aside from these general causes of private-land misuses, there are the causes that are worsened when the commons is divided into smaller shares. Division increases the number and length of property boundaries, thus worsening problems related to externalities. Any division of a landscape increases the challenges of coordinating land uses at large spatial scales so as to address problems that can only be remedied at such scales (for example, protecting wildlife populations, controlling excessive drainage or land-cover change, and managing river floodplains).[20] To break a landscape into smaller pieces adds to these problems, making it even more difficult to coordinate activities and lessening the powers of individuals acting alone to achieve good land use, even when they try. Management at the commons level by no means automatically avoids these problems. But action at that scale provides more options. Harms that are externalities in fragmented landscapes can become internalities, thereby better aligning costs and benefits.

The division of an open-access commons into privately controlled shares can diminish overuse; that much is true and useful. But it by no means ensures good land use and can—in not insignificant ways—make that goal more elusive. The difficulties increase when privatization is compared with the option of commons governance by the users acting in concert. In the grazing setting, for instance, herdsmen in an undivided landscape can use rotational methods that allow them to respond flexibly to variations in range conditions over time and give sensitive areas long rests. This option becomes more difficult, even infeasible, when a landscape is cut into small pieces.

The One Solution

In Hardin's view, as noted, the solution to the tragedy was some version of mutual coercion mutually agreed upon designed to impose limits on individual action.[21] A quick read of his article, though, can give one the sense that he had two solutions in mind, as to some extent he did—government ownership with regulatory control and the private property alternative. For many readers, these came across as solutions that differed in kind, and lots of them much favored the latter.[22]

In fact, however, the private-property approach is merely a form of mutual coercion mutually agreed upon, and not necessarily much different from overtly regulatory approaches. To see this point one has to pause to consider how property arises and how it operates over time, revisiting points covered in chapter 5.

Mythology aside, private property, as we have seen, is entirely a social creation.[23] It arises when a group of people agree among themselves in some way—perhaps democratically but perhaps with elites giving orders—to divide up uses of a landscape in some fashion and to allocate use rights to individuals and families or other groups. Property is a highly flexible institution in that widely varied things can be subject to ownership with rights and limits of ownership that vary just as widely. To create a property scheme, then, a law-making community needs to make key decisions, especially on the basics: what can be owned, how rights will be defined, how norms will be enforced, and what powers lawmakers will retain to change rules over time and to reclaim property (expropriate it) for the common good.[24] Property rights can give owners exclusive or near-exclusive rights to use particular, bounded geographic places. Or they can vest owners with specific rights to use particular places while other owners enjoy rights to use the same places in different ways. Property rights can vary greatly in duration, and owners may or may not gain rights to transfer their property to new owners or to shift to different uses. In some way, property law needs to make rights available to the first owners—to allocate the property—and the possible allocation methods are numerous.[25]

These widely varied options to create private property are usefully compared with what is commonly thought of as the opposing option, government ownership with regulatory control.[26] This latter option also involves government coming up with a legal scheme to control use of the landscape. To that extent, the two options are the same. A government-run system could involve use of the landscape by people working for the government itself or for community members collectively. This option, if chosen, would differ noticeably from many of the private-property variants. More likely, though, a government would retain control of the landscape but make tailored rights to use it (or parts of it) available to private actors, much as on federal lands in the United States today where grazing, mining, and timber harvesting are all done by private actors. When this latter approach is selected, the differences between the private-property and government-regulation options narrow further. In both instances, laws set the terms of use rights. In both instances, systems are backed by state enforcement powers. In both instances, private actors have legally secured use rights and they are the ones who engage in the profit-seeking land uses. At that point, perhaps the main difference is in the flexibility to change land uses; the private landowner has more of it, at least at the small scale, holders of individual use-rights have nearly none, and public-lands managers can make changes only slowly.

When the various land-governance options are all (or a sufficient number and variety of them) set out, they form not two categories of public property and private property but something like an unbroken continuum of rights and land-use arrangements, varying (on one end) from public land used only by government and off-limits to private actors (a sensitive defense installation, for instance), to private land in a remote rural location with nearly no land-use controls and few spillover effects. In between these poles would be lands with varied mixes of private and public control: with variations on the scope and nature of private use rights and, on the public side, in the retained powers of government over time to revise the terms of the use rights and redirect resources to different uses and users.

The Challenges of Avoiding the Tragedy

Hardin cannot be faulted for the idea that the private-property remedy somehow differs in kind rather than degree from the regulatory option. He might be faulted, though, for not saying more about the grave difficulties involved in setting up any coercive regime, one that allows intensive uses of a landscape but keeps the uses within proper bounds—a regime that allows full use but somehow forestalls abuse. In reality, the work involved in this law-making or norm-creating is quite considerable. It is hardly enough simply to draw lines on a map, dividing a landscape into shares. Far more labor is needed to craft workable rules and to enforce them over time. The challenges in doing this are many, and they arise in every land and resource setting. They can be acute in settings where nature itself is highly dynamic and when (as often) actions by one person directly affect both other people and the resource itself, understood in ecological context. They can be acute also, if not insurmountable, in a culture not inclined to embrace limits and when too many people are not inclined to cooperate.

As an initial matter, it seems obligatory in any setting to start with the basic distinction between the legitimate use of nature and the misuse of it. How should lawmakers differentiate the two? This is, as noted, a normative task, and a challenging one when done wisely and morally, when it takes into account, as it should, the full range of factors relevant to it.[27]

This line-drawing is a necessary element in any rights-allocation system because, for reasons mentioned, market forces and human nature are not such that a community can expect individual users voluntarily to stay on the right side of the use/abuse line. If a community really wants to halt misuse, then it needs to tailor use rights under any scheme so as to allow legitimate uses and disallow all misuses (Hardin's main point). A legal scheme need not rely solely on direct prohibitions of misuse, particularly in the case of land- or resource-use harms that entail what might be termed carrying-capacity harms: actions that are harmful only when too many people engage in them.[28] Lawmakers might instead use other, less peremptory management tools, perhaps involving economic incentives, perhaps instead involving efforts to nurture and strengthen social norms.[29] Still, a line must be drawn, which means, somehow, doing the work of drawing it. This work is likely to call for considerable ecological knowledge, as well as mature thought on the various relevant normative factors. The need for ecological knowledge in turn can require extensive scientific study.

With this line-drawing done (for the time being), the next step is to craft use rights so that they allow land and resource uses consistent with it. In the case of highly varied landscapes, this will likely mean tailoring the use rights to nature itself so as to allow uses that are ecologically sound and to limit those that are not. The guiding principle here would be rights that are based on nature in the sense that they take natural features into account.[30] Opposed to this—to highlight the point—are property rights defined abstractly, as with the hypothetical Greenacre or Blackacre: rights defined with no regard for natural features or for the effects of a given resource use in a particular place.[31] This work too may be quite challenging. And, again, it needs to be done under any coercive option, whether termed private property or government control. It needs to be done, that is, if the goal (as assumed) is to avoid abuse entirely and if the community of users is not so close-knit, and so well guided internally, that individual owners (*contra* Hardin's tale) can be counted on as individuals to avoid bad actions on their own.

Making this tailoring job more difficult is nature's dynamism and the fact that human actions themselves inevitably change the parts of nature being used.[32] Change in physical conditions in turn means, in dialectical fashion, changes in the future uses that will be permissible under the new conditions. Nature's dynamism poses challenges in addition because it gives participants (so the evidence shows) a further excuse for resisting limits. Also shifting along with nature will be prevailing ideas about good land use—about the line drawn between use and abuse—shifts that could occur because of new factual knowledge but could also occur due to evolving communal needs and values.[33] The factors interact in a kind of multifactor dynamism, one that can prove especially knotty in the case of biological resources (fish, for instance) where populations are subject to wide natural variations and where the capture of one species can distort populations of other species.[34] It can prove knotty also due to variations in weather patterns—drought, flooding, extreme temperatures—and to natural disturbance regimes such as fire.

In some way lawmakers charged with the task of crafting private use rights need to take this dynamism into account. They could do so by erring on the side of great caution, by allowing year in and year out only those limited land uses that would respect the land's ecological health under any and all foreseeable landscape conditions. That approach, however, is likely to be rejected as much too cautious. The alternative is to allow more intensive use levels but to embed into the private-rights scheme some mechanism to curtail use levels when needed to respond to these dynamic factors. A fishery is again a good example: harvest levels need to vary with natural fluctuations in target-species populations. Various methods could be used to build in this responsiveness. All of them, though, require ongoing monitoring, data collection, and adjustments to use levels, just as they require a cultural willingness to embrace limits. All of them require ongoing involvement in managing use levels at the landscape (or fishery) level. It is not enough under any option for lawmakers to set usage levels once and for all and then walk away. In short, a property rights approach requires ongoing manipulation, just as does the regulatory option, if it is really going to avert overuse.

To get to this point is to see that the property rights and regulatory approaches may not be all that different. Of course they could be quite different in a particular setting. Lawmakers taking a property rights approach could define private use rights in clear, unchanging terms, without making them responsive in any way to shifting natural conditions. But by doing that, they give up the power to protect the resource against overuse. Their lawmaking effort might still reduce Hardin's tragedy but would hardly eliminate it. To gain full protection, the property rights approach would need to take shapes that rather closely resemble a well-designed regulatory system, perhaps quite closely.

In the end, the work required to create a property rights system might be not all that much less or different from the work needed to run a successful regulatory system.[35] Both require extensive, sensitive efforts by lawmakers that attend closely to ecological facts. Both require a spirit of cooperation and concern for the common good. Both require lawmakers to resist pressures by user groups to authorize higher use levels than are consistent with the avoidance of overuse. Can lawmakers be counted on to do this? Are they likely to resist pressures to set permissible use levels too high? Probably not, most would say; surely not, others would contend. Biased lawmaking is indeed a danger if not a high probability. But it is a danger, we need to see, under *both* the property rights and the regulatory approaches. The property rights approach might lessen this problem of ignoring limits but the problem hardly goes away.

In this light, is there reason why we might have greater confidence in lawmakers when they are specifying the elements of property rights (wearing one hat) than when they are crafting more overtly regulatory tools (wearing a different but similar hat)? Is it sensible to blast regulators as incompetent but then expect them (or their statute-writing colleagues) to do vastly better when it comes to specifying property rights?

Root Causes, Again

In Hardin's simple story, the grazers overuse the pasture because they want to increase their individual short-term profits. That motive is not merely the primary one; it is, apparently, the only one, with no other factors in supporting roles.[36] Hardin's explanation has obvious merit to it; profit is patently a strong lure. But as a summary of the root causes of degradation—of the factors and reasons why humans misuse nature—it is nowhere near complete. The factors at work could well be quite numerous, and they are when we consider many settings. We need to identify and trace these other causal factors before Hardin's tale can stand as more than a crude explanation why people act as they do.

What other factors might be at work in a grazing story such as Hardin's?[37]

One possibility, already mentioned, is that a grazer might act in ignorance, whether about the carrying capacity of the land or about the numbers and grazing plans of fellow grazers. Hardin looks down on his created scene from above, knowing all. Grazers on the ground might know a lot less. Related to this, the grazer might disagree about what qualifies as overgrazing— or, more generally, about where to draw the line between legitimate use and abuse.

As for the selfish drive for short-term profits, an individual grazer might be inclined to resist it. But what is the consequence for an individual grazer who refrains from adding more cows? If the consequence is that the grazer thereby improves pasture conditions, then the grazer might hold back. But to bring about Hardin's tragedy we do not need *every* grazer to be selfish. Not every grazer needs to make the calculation that Hardin presumes. We probably need only a few grazers, maybe just one. One aggressive grazer alone might add enough animals to degrade the entire pasture.[38] When decisions are made collectively, a majority favoring restraint might carry the day; when decisions are made individually, land protection may require not just that everyone exercise restraint, but that they have confidence that their neighbors will do so as well.

This possibility gives reason to modify Hardin's simplistic story. A conscientious grazer who wants to avoid overgrazing and is willing to do his part, willing to join in a system of mutual restraint, is put in a bind. Yes, the grazer can individually refrain from adding animals. But if the tragedy is going to take place anyway, if other grazers can ruin the land without his help, what is the benefit? There is no longer a tradeoff of either higher short-term profits and a degraded pasture or lower profits and a healthier pasture. The landscape will decline in any event. So what is the point of holding back?

A grazer who appraises the situation this way and then, perhaps with resignation, goes ahead to add cattle, is not acting selfishly, not in any full sense. His situation is akin to that of the nature lover who wants to see a wilderness area left untouched but has no power to keep others from altering it. The person might then go ahead and invade the wilderness, despite wanting it protected, simply because others will do so anyway. To refrain might be virtuous in the sense of reflecting good character. But it is not ethically required under any ethical system based either on rights or consequences, not when the harm will happen in any event.[39]

To these possible motivations for overgrazing we can quickly add others, many of which also explain abusive practices by private landowners.[40] A grazer could simply be short-sighted and not care about the future or could use a high discount rate that reduces the weight given to future harms when compared with present gains. The simple fact that the grazer is willing to push the land to its limits to meet human wants (rather than acting more cautiously), and the fact that he views land and animals as objects of manipulation (and not, in the case of the cattle, as fellow creatures), also play explanatory roles in the chain of events. As best we can tell, Hardin's grazers view their landscape simply as a place to raise cattle; it is not a multiple-use land that might produce varied benefits.

As for the always-important economics, we know nothing about the grazers and how they are faring. Might they need extra money to feed hungry families or to pay impatient tax collectors? Knowing as little as we do about the context, it is presumptuous to guess motives and judge conduct. Starving people do not steal food because they are selfish.

Another key part of this whole tragic saga, again implicit in it but not highlighted, has to do with the market in which these grazers are apparently embedded. A grazer who raises ever more cattle to make money is likely raising them to sell. To note the presence of a market and the chance to make money in it is to bring in another key causal factor. Take the market away and the overgrazing might even end, at least if the grazing pasture is more than adequate to meet the direct needs of the people who live there.[41] Take away the market and its unlimited demand—as well as any commands from on high that they raise cattle to export—and the grazers may be more willing to acknowledge the wisdom of restraint.

Then we have the question that has loomed above Hardin's tale from the beginning, the question that should have been, for decades now, at the center of discussions about Hardin's tragedy. If his grazers could have come together collectively to develop a management plan, benefiting all of them, why did they not do it? Why did they continue competing as autonomous individuals rather than cooperating for shared gain?

In an important sense, the real cause of the tragedy is precisely this: the failure of the grazers to communicate, to get together, and to act sensibly. So why did they fail to cooperate?

To raise this issue is to highlight yet again how ignorant we are about this simple grazing commons and to see how this ignorance can (and does) readily lead to mistaken assumptions. As possible answers to these questions—as explanations why competing grazers might fail to communicate and work together—it is easy to round up some likely suspects.[42]

- The bad communication could stem from differences among the grazers themselves, differences in language, ethnicity, religious, race, and so on.
- It could be due to membership in different political or tribal groups that do not get along.
- The grazing region could be (probably is) a borderland where group competition plays out; historically these are the grazing regions most often misused.
- Perhaps the prevailing political and social culture is weakened by oppression so that the grazers, fearful of informers, secret police, or organized crime, refrain from talking to one another.
- If grazers are taxed for each animal they own they may just refuse to talk about their herd sizes and to question others about theirs.
- Finally, the grazers might simply feel helpless and resigned. They may assume that any arrangement they concoct will collapse under pressure from outside powers—from powerful cattle buyers (global corporations?), for instance, driven by neoliberal agendas.[43]

All of these possibilities need to be placed on the table as causes of landscape tragedies. Tellingly, none of them appears in Hardin's simple story. For us to interject any of them requires that we add facts to his tale, thereby making it more plausible. There is one big causal factor, though, that is present in Hardin's narrative without any need for us as readers to add complexity. This is the inclination of his fictional characters simply to look after themselves as individuals, ignoring everyone else.

Each grazer, Hardin tells us, responds solely to the costs and benefits he incurs as an individual. It is a culture of go-it-alone, a culture in which individuals look at their neighbors as competitors, not colleagues. As Chris Williams has pointed out, Hardin presents this motive as "a transhistorical fact," as if it identified an unwavering human trait. Hardin, in fact, describes as rational "the very thing that traditional herdsmen and peasants sharing 'common' lands historically avoid."[44]

For the grazers in Hardin's story to act as they should have, for them to have come together and developed a pasture-management scheme, they needed to live and work within a culture that expected and nurtured cooperation.[45] They required a home culture in which people acted as citizens or community members as well as market participants. Further, they required a culture—if they were to sustain healthy pastures—that generated and shared ecological wisdom about landscapes and that included vocabulary suitable for talking about it. The ideal of land health, or something like it, would need to stand as a matter of common concern, suitable for common resolution. For that to happen, the people would need an awareness of nature's functioning and a willingness to restrain themselves to respect that functioning. Finally, the culture would need to view collective action as a legitimate way for people to pursue their aims. It could not be a culture in which people resist all collective governance. It could not be a culture in which land abusers could ward off critics by raising high the shield of individual liberty.[46]

A land-respecting culture can certainly embrace private property.[47] But it cannot view private rights as so strong and inflexible that cooperation among owners occurs only by unanimous agreement. If private rights are inflexible, if grazers can halt overgrazing only when everyone agrees, then a single dissenter can frustrate everything. In short, strong private rights—whether or not phrased in property terms—can quickly cause ruin, even when nearly everyone wants to avoid it. And it is hardly fair to say that people who value land health simply need to pay abusers enough to get them to stop. Why should they bear such a burden?[48]

The larger issue here, to reiterate, has to do with the root causes of tragedy-inducing behavior. This tragedy, like all environmental problems, is brought on by human action. The search to learn why people misbehave should thus be central to any inquiry into our environmental plight. Hardin cannot be faulted much for largely ignoring the issue; for assuming simplistically that overgrazing stems from one factor. But commentary on it, and certainly commentary on our environmental plight generally, has been much weakened by a failing to put this issue front and center. Important exceptions, as we have seen, include Leopold, Berry, Orr, and, most recently, Pope Francis. Yes, when a market exists and people desire to get ahead they take steps that make them money. To that extent Hardin's tale is well grounded. But in a culture that honors land health, one that facilitates discussion on key normative issues, that encourages cooperation as well as competition, and that views government as the people's agent for shared work: in such a culture, the profit motive might fade in significance.[49]

Today's culture includes and gives priority to cultural elements that are root causes of ecological degradation. It features cultural elements that foster land-use tragedies while frustrating efforts to halt them. Particularly in its liberty-loving, antigovernment forms, modern culture undercuts efforts to see landscapes as ecological wholes and to think normatively about their conditions and prospects. It holds high the rights of individuals to go it alone like Hardin's grazers so long as they respect the rights of others to go it alone also. As Hardin illustrated, this is a culture that leads readily if not inexorably to disaster. Responsible libertarians, to be sure, call for compliance with laws, property law especially. But the culture they press forward is a culture that frustrates good lawmaking, including the lawmaking required to come up with new well-crafted property rights.

When mutual coercion is the only solution to avoid tragedy, as it is in setting after setting, a culture that despises and resists such coercion—a culture that presses hard to keep it from happening and, when it does take place, views the lawmaking arena as simply another venue for individuals as such to get ahead—is a culture that is driven to ruin. The work of mutual coercion, of crafting collective management schemes, is difficult even when all participants support it and bring to the table their best inner selves. When many of them instead do all that they can to undercut it and/or to manipulate the rulemaking to benefit themselves—when the culture they embrace views such self-seeking behavior as simply vigorous competition—then the path to hope becomes yet darker and more treacherous.

Naming the Tragedy

The title Garrett Hardin gave to his essay did not really name the tragedy that took place in his grazing landscape (or in his overpopulated world). Rather, his title seemed to identify *where* the tragedy occurred. But Hardin does imply that his tragedy unfolds because the pasture is not governed by an effective land-management regime, by some version of mutual coercion mutually agreed upon. It is the absence of good laws that is the problem, and a new legal arrangement, he tells us, is the only solution. In that light, his title identifies not just the place of the tragedy but also its cause. Government (assuming it exists) is to blame for not adequately controlling the self-seeking individuals.

One might readily object to this shifting of causal responsibility. A well-constructed legal regime is the *solution* to the problem, not its *cause*. The grazers bring on the overgrazing themselves. The true causes of the tragedy are the forces and factors that lead the grazers to act as they do rather than in other, more responsible ways. Government is not to blame, nor, really, is the commons itself, even if we could somehow view nature as an active agent in the unfolding drama. When the commons is used by people who collectively want to use it well and who have the knowledge, skills, and opportunity to do so, then (and only then) does the outcome brighten.

If we want an apt name for Hardin's narrative, then, we need to look in another direction. An apt title would be one that summed up, in a few words, the key cause of the misbehavior. As for that, the possible titles are many, and it might prove helpful to debate their relative merits as an avenue for probing the likely causal factors.

One place to turn for a new name is to Bernard Mandeville's eighteenth-century classic work, much liked by libertarians and market enthusiasts, *The Fable of the Bees: Or Private Vices, Publick Benefits.*[50] Mandeville was an early advocate of the view, later connected with Adam Smith, that linked energetic money-making efforts to economic growth that benefited the public generally. Mandeville's thinking, of course, is best understood in the context of his worldview and its distinct (and now ignored) assumptions about a binding moral order. But we can put his age aside and take inspiration from his title. Hardin's tale might be termed "The Fable of the Cows: Or Private Vices, Public Decline."

As noted, the overgrazing in Hardin's tale is also closely linked, essentially so perhaps, to the unmentioned market in cattle. Take away that market, take away the grazers' tendency to view cattle as capital assets, and the overgrazing would diminish, if not end. We could thus consider, as a second option for the title, "The Tragedy of Market Capitalism."[51] If selected, this phrasing might prompt readers to see links between tragedies of the type Hardin described and other instances of wastefulness and degradation brought on by the vigorous pursuit of profits. Detroit, Michigan, could supply ready facts for several illustrations. Variations of this title could draw attention to particular elements of the market, to its commodification of nature, for instance, or to the short-term perspective that it invites and rewards. In a capitalist system, market participants are expected if not compelled to cut costs, by externalizing harms and other means. They are pushed also to expand and to focus on short-term profits, calculated in ways that ignore much environmental degradation. Absent a long-term perspective they are encouraged to deny the reality of limits and to resist those who call for them. They are encouraged to embrace the mythology of nature's unlimited abundance—the "theory of the green light," as historian Donald Worster has recently termed it.[52]

One might be tempted as a further option to choose a title that gets at the selfish individualism involved in Hardin's tragedy, at what might be termed the cussed individualism. But we need to be careful with this phrasing, tempting though it is. In Hardin's tale, the grazers as individuals would have been better off had they worked together. The problem—in this setting, though not in all—was thus not the grazers' presumed desire to flourish economically. It was, more exactly, their failure to pursue that goal by working in tandem. An accurate name would need to get at this particular aspect of individualism, the "go-it-alone" version of individualism, the kind of individualism that fails to recognize how a responsible person plays many social roles with expectations of good conduct attached to them.

Two years before Hardin's article came out, the economist Alfred Kahn wrote a rather similar article that highlighted the same problem. Kahn's article drew less popular attention, very possibly, one might guess, because he phrased the issue in a way that many readers liked far less. Kahn's essay was entitled "The Tyranny of Small Decisions."[53] It highlighted how individuals who made decisions in isolation, even acting in economically rational ways, could drag down the communities to which they belonged. When people stood apart, when they chose or were compelled to act in isolation, their individual "small" decisions brought harm to the whole. This title was later picked up by an ecologist, William Odum, who used it to help explain the root causes of ecological degradation.[54]

Kahn's title was far more apt than Hardin's in terms of highlighting the cause of the bad outcome. It pointed a finger more directly at the individual grazers—it was their decisions that brought on the harm, not any failing of government—and it emphasized that the problem had to do with their isolation or autonomy. So long as the grazers acted alone, not together, their decisions were bad ones. Kahn's title, then, could be recycled. Hardin's tale is very much one in which small decisions exerted destructive power; it too shows the tyranny that they could bring on.

Small decisions become more likely when a landscape is divided into many small pieces and when management of the pieces is turned over to individual owners. As Hardin points out, the division of the landscape into private shares might well diminish the harms that unfold in a pasture that is completely uncontrolled.[55] Privatization is a step in a good direction. But it is only a step and it is unlikely to sustain the landscape's long-term productivity. For reasons mentioned, fragmentation can leave key problems unaddressed. And it can make many forms of good land use more difficult. To get at these problems we might then select as a new title something like "The Tragedy of Fragmentation."[56] Like "The Tragedy of Market Capitalism," this title also has broader usage. It can refer, not just to the fragmentation of landscapes in physical senses, but to other forms of fragmentation: to views of nature as a collection of pieces and parts rather than integrated ecological systems; to views of humans as distinct autonomous beings detached from social bonds and surrounding communities; to the disconnection of the present generation from past and future ones; and to the division of life forms into distinct categories with widely varied moral value. As a title, "The Tragedy of Fragmentation" more directly engages issues of ontology. It challenges views of being that emphasize the organic parts and discount the wholes that they help form together with the emergent properties that arise only at such higher levels.[57]

In the end, however, perhaps the most apt new title for Hardin's tale would be one that points as distinctly as possible at the reasons why the grazers might fail to get together to cooperate. Cooperation alone is certainly not enough to avoid bad outcomes; the grazers could cooperate to exploit the commons with zest, take their profits, and run. Big decisions can be worse than small ones. Cooperation thus needs to work hand in hand with sound values, with extensive ecological knowledge, and with a form of humility that reflects human ignorance about changing nature. But the cooperation nonetheless remains the key. The grazers simply must get together, plain and simple. They must have the sense that their lives and fates are intertwined, and that they can achieve better results working together than working in competition. If external forces keep the people apart, then we need to point to those forces and make clear their costly consequences. But if as usual (one suspects) the grazers are kept apart because of reasons internal to their membership, because of the flawed culture in which they operate, then the name for the tragedy should point in their direction.

In some way, a name that gets at this root cause should hold high the need for the grazers to think of themselves as bound together in a community, or more exactly, as embedded within at least two communities, the social and the ecological. Practically speaking, a good grazing arrangement will respect nature as an interconnected, interdependent community of life. Humans will appear embedded in that community and ultimately dependent on its long-term health. In social terms, the grazers need to see similarly that they belong to human social networks and depend on the strength and good content of those networks. Fully good land use is possible only when the grazers recognize these communities and, having recognized them, work to sustain them. In that sense, we can rightly view the health of a grazing region as a condition or property that emerges only at the community level, only when both the land community and the human social community are healthy and functioning well.[58] Good land use is thus an emergent property, generated at the community level and only at that level. Seeing this, we might then choose as our title "The Tragedy of Weak Communities" or "The Tragedy of Incomplete Communities."

These various names each have appeal. Perhaps they could be used somehow in tandem. Yet one more name can usefully go on this list, a title that seems most apt in the modern age, most apt in a culture guided by the kind of libertarian, free-market ideology that has gained such prominence in the United States of late.

The liberal trajectory of the Western world over the past three centuries has brought substantial gains in many realms, of that there is little doubt. In economic realms it has fostered substantial increases in economic productivity, sizeable enough overall to exceed in value the considerable associated costs. In social realms liberalism in its more progressive forms has brought great gains for the downtrodden and unfairly burdened. In political realms it has spread political power through increases in suffrage. Driving much of this has been a commitment to rationalism and empirically grounded science along with a wide-ranging willingness to challenge and revise inherited understandings. Providing the polestar has been the liberated, rights-bearing, individual, increasingly free to develop and embrace values as she sees fit. This liberal social thought did not have to carry over into the ways people viewed nature. But to some extent it did, Nature, too, came to seem more fragmented, dynamic, and competition based, just like the modern human economy. This view of nature in turn made it easier to relax traditional limits on economic enterprise.

When we revisit Hardin's tale in light of this long-term cultural trajectory, we can see that the bad-acting grazers that he describes are in fact familiar characters. They are exemplars of the autonomous, self-serving competitive individual, most familiar in today's marketplace. Setting to one side possible external causes (oppression, corruption, or political instability, for instance), it was this full-bore liberation of the individual, particularly in the economic realm, that brought destruction to the landscape. Hardin's grazers showed no signs of allegiance to their social or natural communities. Freed of communal restraints and viewing fellow grazers as competitors the liberated individual grazers could proceed as they saw fit, pursuing short-term profits and degrading their natural home. Thus we might consider, as a final title, "The Tragedy of Liberation Taken Too Far."

Thinking, Talking, and Culture

The studies and explorations developed in the above chapters have returned again and again to common topics and observations. Before taking up what might be done today to foster reform, to improve how we dwell in our common home, it should help to reprise these topics and add a bit, both to draw them together and to get clearer on why environmental reform efforts today, if they are to gain real force, need to shift focus.

Good Land Use

A central theme throughout this inquiry has been our need to think more deliberately and deeply about how we ought to live on land, about how we might best distinguish between the legitimate use of nature and the abuse of it. To draw such a line, even when it remains subject to revision, is to have in view an overall goal for conservation efforts. It is to have at hand a way to identify land-use practices that are problematic and to spot the human actions causing them. Vague ideas about sustainability are not enough and even more specific ones—framed in terms of ecosystem services—have troublesome gaps. Today's disagreements about environmental problems—whether they exist, how serious they are, whether they are worth addressing—often arise out of underlying, unstated disagreements about the proper normative standard for evaluating our conduct. Development pressures can fragment a forest with dozens of new home sites and their accompanying roads and utility lines. Civic boosters can view this as progress. Environmental organizations might interpret the landscape changes more darkly, citing the habitat fragmentation, changes in hydrological cycles, and increased vehicle traffic and related pollution. The two sides, to be sure, may disagree on some of the ecological facts. But they likely disagree even more on the proper standard for evaluating the facts. And they do not, in any meaningful way, talk about that evaluative standard. They may all pledge support of sustainability, yet disagree so widely on what sustainability means that their agreement counts for little. Any discussion, if and when begun, could quickly shift ground to the issue of private property rights, understood in individualistic terms. At that point, hope of serious talk on the common good can disappear behind claims of individual liberty.

A vision of good land use was central to Aldo Leopold's thought in his mature years, and he had just begun, before his death, pushing conservation colleagues to rally around it. Leopold, as we explored, used the term "land health" as a short-hand summary for his ecologically grounded normative vision. All conservation should aim toward it, he claimed. Wendell Berry in his writings has paid less attention to ecological functioning, as we have seen, but has still held high as the one value, the overriding priority, the health of the land as a whole. Beneath the surface one finds the same basic idea in the writings of David Orr, in the visions he casts to guide his ambitious efforts at overall design. Pope Francis in his encyclical identifies an array of normative considerations that are essential in constructing a guiding vision of flourishing life, and he has urged us to put them to use.

A vision of good land use, as it came together, would almost certainly emphasize the need to protect the basic ecological functioning of all lands, which includes keeping and building soil, maintaining natural fertility cycles, protecting and restoring as many species as possible, and reversing many changes to hydrologic flows: reconnecting rivers to their floodplains, reducing artificial drainage, limiting irrigation to the most highly valued crops, removing countless reservoirs, and more. It would similarly give primacy to protecting well-chosen wild or semi-wild lands in all large landscapes. As important, the talk about good land use needs to raise big possibilities. It is by no means clear, to offer one example, that all the locks and dams on our inland water ways bring any net gain; they certainly come at vast ecological costs. The grain and coal carried by barges on the Mississippi and its tributaries could all travel by rail. A very sensible goal might be to return the Mississippi River (or most of it) over the next 50 years to something like its condition 150 years ago. Similarly, the farmlands of the Midwest call out for radical change in land-use types and patterns, shifting to far greater crop diversity (including tree crops and other perennials) and greater interspersion of land uses, perhaps something along the lines outlined in Mark Shepard's visionary *Restoration Agriculture*.[1]

Guiding the work and talk about good land use ought to be visions of how our natural homes can be better all around, including healthier. But the talk must also include an overt recognition of nature's limits and a confession that we now push nature much too hard and far. It must similarly include a recognition that our changes to nature routinely create unexpected harms, which then require further attention and costs. The talk of limits should not predominate; it is likely too focused on what we must give up. Yet limits are real, even as they depend in their details on our normative values and hopes. To see nature clearly is to keep limits always in sight.

Some such land-use vision is needed for the work of recalibrating private property rights in nature if they are to enjoin landowners to use their lands and resources in communally good ways. Private property is best understood as a tool used by a lawmaking community to foster its shared aims. As many writers on the institution have insisted (including such liberal founders as Jeremy Bentham and John Stuart Mill), the powers of landowners are derivative of the common good and should extend only so far as consistent with it. Although using different wording, Pope Francis strongly concurs. The good use of land is surely a central component of the overall common good. Well-structured private property rights should help promote it.

Wilderness protection is similarly linked—it ought to be—to a vision of good land use at large spatial scales. A prime benefit of wild places left largely untouched is that they help sustain the ecological health of their surrounding landscapes and the flourishing of life within it. It is hard to know how they might do that, it is hard to explain this particular wilderness benefit publicly and clearly, without having in hand a reasonably clear measure of good land use. When good land use includes, as it likely should, the protection and rebuilding of wild species, then wilderness-type lands become essential. Similarly, a definition of good land use is needed to evaluate uses of landscapes that people in some sense work in common. A tragedy of the commons unfolds, pretty much by definition, whenever overall uses of a landscape cross the line between legitimate use and abuse. A measure of good land use is needed to evaluate common landscapes. It is needed similarly as a goal when developing mutually binding coercive rules to help end the tragedy.

Aldo Leopold spent much of his professional life piecing together his normative vision of land health. It is challenging work, as Leopold knew and as we need to see, calling for engagement on the full range of relevant normative factors. Writings on ecological sustainability offer useful insights, as do the related writings on ecosystem services. Human needs and hopes will feature prominently in any well-crafted standard. So will a recognition of the limits on what humans know and can know and thus of the wisdom of acting cautiously and leaving room for second chances. A long-term view, one that looks forward to future generations, surely ought to characterize the work, as Pope Francis has insisted. So should reflection on the moral status of other living creatures, as species, biotic communities, and individual beings—the recognition that our common home is morally infused. Given how uses of nature by one person can affect other people, social justice considerations are plainly relevant, particularly the need to share the planet equitably. As we have seen, social justice and the needs of the poor figure prominently in the thought of Pope Francis, who stresses how the root causes of ecological decline overlap with those of social inequity.

In some way, people need to come together, better than they have, to embrace at least a basic vision of good land use. To do that, they need to recognize that science alone cannot get the job done, even as it supplies essential knowledge. We cannot soundly jump from the "is" of nature as described by scientists to the "ought" of good land use. The line-drawing work is normative at root, calling for choices about right and wrong, wise and foolish, and moral and immoral. To do the work, it must seem right for people to engage important normative issues in the public arena, and not view them as matters for individuals to address as they see fit in their private lives. This means, as taken up below, rather significant changes in the ways we commonly understand individual autonomy.

Root Causes

The issue of root causes similarly runs through the various inquiries in this book, both on the surface and beneath it. Wendell Berry from his earliest writings has seen the origins of bad land use in bad aspects of culture. He has portrayed and probed those bad aspects—as well as good counterpoints—in book after book. A recurring cultural flaw for Berry has been a shared inattentiveness to connections: a failure to see well enough how people are linked to one another and how the present generation is linked to past and future ones; a failure to appreciate the often invisible forces of nature that come together in poorly known ways to give rise to the miracle of new life. Like Aldo Leopold before him, David Orr has returned again and again to the need for cultural change, for greater awareness of ecological realities and interconnections, for the need to work with nature, taking advantage of its evolved wisdom, even while using the best science and technology to meet human needs with minimum ecological impact. Pope Francis is equally insistent that we abuse nature and one another because of defects in our culture, and that true reform needs to center on cultural change.

The message recurs: We need to see nature anew, to recover an understanding of it in terms that, if not overtly holistic, at least recognize and value the links and processes that orchestrate natural systems and account for its operations and productivity. To do so is to counter the costly tendency to fragment nature unduly and to value its parts as parts, often through the market. When nature is seen instead as an interconnected system it becomes easier to talk about it as our common home and to exhibit for it gratitude if not spiritual reverence.

Many of the cultural root causes of degradation are linked to traits of the dominant liberal tradition, particularly in its less-responsible libertarian forms. They have to do with the tendency to raise high the individual human and to liberate him or her from communal constraints. The dark aspects of this type of individualism appear perhaps most vividly in tales such as the one recounted by Garrett Hardin. There we see the individual going it alone, promoting his self interest in isolation from the welfare of others and pushing land well beyond its carrying capacity. Hardin's tale and others like it need revisiting and studying again, looking particularly for the factors and forces—social, cultural, and economic—that lead to misuses of nature and that inhibit collective action. It is unlikely that we can avoid such land-use tragedies without confronting their root causes in culture. We likely need to confront them also if we are to redefine property rights so as to allow owners to prosper and flourish, as individuals and community members, while also ensuring that land uses are ecologically sound.

Ecological ignorance surely plays a role in the misuse of lands. But it is telling that, in the writings of Leopold, Berry, Orr, and Pope Francis, one sees little emphasis on public education aimed simply at disseminating facts. Hardin's tragedy, so far as we can tell, did not unfold because individual grazers lacked particular facts. Similarly, the roadblocks to needed reforms in property rights are grounded far more in individualistic values and presumptions than they are in any facts, about nature or anything else. The public's views on nature, and the human place in it, are in need of realignment.

Beyond Autonomy and Arrogance

The path to hope, then, is one that will necessarily entail challenges to entrenched ways of seeing and interpreting the world. It will need to bring reforms in the components of modern culture that fuel misuses of nature and that, in doing so, make the harms seem like necessary byproducts. Down the path of hope must appear a reformed culture, one that understands the human predicament in modified ways, a culture in which better land uses seem both logical and emotionally satisfying. Overall, a reformed culture will show greater respect for ethical holism in many forms and settings, greater moral respect for interconnections and for natural and social communities as such.

Medieval culture, in its own ways, honored human exceptionalism, basing its claims on biblical texts of human superiority. But humans then were embedded in a larger structured order. If all other living creatures were below them, the angels and the Trinitarian divinity in the Christian version ranked higher still. Humans were by no means in charge. It was the Renaissance era when people began seriously to think of man as the measure (an ancient Greek idea) and only in the seventeenth century did human reason seem potent and reliable enough to generate truths apart from religious texts and traditions. From there humankind rose steadily upward in importance, albeit with many steps sideways and downward and with relentless resistance. With the Enlightenment and revolutions, and then with the overt emergence of economic and political liberalism, humankind in individual form gained yet further prominence in Western thought. The work involved in this—weakening communal bonds and traditional norms—was never easy. But the forward motion nonetheless continued, even with setbacks as large as World War I and the horrors of German nationalism and Soviet and Chinese ideologies.

Grand historical surveys of this type, of course, are easily countered with contrary evidence, and such evidence over the centuries has been abundant. Nonetheless, it is undeniable that ideas of human moral value and individual rights have greater power today than ever before. The mysteries and spirits once embedded in nature have been pushed aside, either gone entirely or recognized only in private realms with little influence on public policies. The market-based view of nature as commodities has correspondingly gained great influence, even as the romantic allures of nature remain strong in literature and private lives. On contentious public issues, facts and reason are expected to play the central roles while subjective preferences are left for individuals to act upon as they see fit. Individuals as such are the ones who hold value, not families, neighborhoods, tribes, catchment basins, biotic communities or other collectives. This individual-centered persuasion—this liberal persuasion, as the term was originally used—might not stand as dominant as a few historians once claimed (Louis Hartz, most prominently).[2] But our shared tendency these days is to understand the social order as a collection of rights-bearing individuals and to assume that, all things equal, greater autonomy for each individual will bring good results. Greater individual liberty has indeed brought real advances, to be sure, as many civil rights campaigns attest. But as a line of thinking and valuing it collides with the realities of a natural order based on interconnection and interdependence. It collides with the recognition that how one person lives and acts inevitably affects others, and that good land-use patterns can only arise through collective work. Such work, in turn, requires that we see ourselves, as Leopold, Pope Francis, and others have urged, as shared inhabitants of a splendidly rich, highly complex natural home.

The reform path that seeks to redirect this dominant trajectory needs to reconsider our presumed human exceptionalism, not ending it but modifying it in something like the ways proposed by Pope Francis and others. We have many reasons to value and protect human life and to treat it as special. But as we do this we need to grasp that our presumed moral superiority rubs hard against the physical realities of nature. In the natural order we are living creatures, similar to all others. We have greater powers to alter nature but still remain subject to its many functional processes and our knowledge about it remains incomplete. As Leopold explained, we are as embedded in the land community and dependent on its long-term health as the earthworm. That reality should constrain our assumptions of being different. Our bedrock faith in individual moral value is most useful and appropriate when deciding how to deal with one another. It is far less pertinent when it comes to our links to nature and how we should interact with it.

Generations of humans have worked and died to push forward claims about human dignity and freedom. It is folly to reject the good fruits of this work. It is not folly, however, to confine these fruits to settings in which they are useful and to steer them away from settings where they are not. It is not folly, when thinking about our common home, to recognize value also in its nonhuman members. It is not folly to embrace laws and norms that constrain the liberties of all to protect the interconnections and natural processes basic to all life.

Among the most troubling aspects of individual autonomy is the too-simplistic view that we should leave individuals as free as possible to select their preferences and values and then to use their preferences to guide their personal lives. It is a cultural ailment that warrants sustained emphasis. Such reasoning honors the individual as such and is embraced because it does. But taken to its logical end, the reasoning gives rise to the misguided ideal of the neutral, night-watchman state. It proposes a government that merely keeps peace and order without taking stances on other normative issues. In fact, there is, on key issues, simply no way for governments to remain neutral. To allow businesses to deceive consumers, for instance, is hardly neutral on the issue of acceptable business practices. Beyond that, many normative goals, land health included, are ones that are well beyond reach when people act only as individuals and voluntary groups. No single grazer can keep the pasture healthy. Given nature's ubiquitous interconnections, all land uses are legitimate public business.

Current laws and policies already display strong normative leanings, quite often slanted to favor businesses, economic elites, industrial farmers and foresters, and fossil fuel producers. To leave these law and policies alone, to strip government of the power to change them, is very far from normatively neutral.

Government writ large—at all levels and including civic society—is the means by which people collectively achieve goals they cannot achieve acting as autonomous individuals. That truth needs recovery, even as we remain mindful of the known dangers of majoritarian rule.

Collective Action

Aldo Leopold had no illusions that land-use laws and regulations could ever be detailed enough, on a parcel-by-parcel basis, to compel landowners to use their lands in good ways. A yearning to live rightly on land, Leopold sensed, had to reside in the bones and hearts of the landowners themselves. It had to be part of the culture that framed their views of nature and that motivated them to embrace their lands as ecologically vibrant, complex natural systems. Leopold had farmers in mind most of all, and he never really translated his ideas to urban and suburban settings. But he had the same message of cultural change for landowners in such places, and indeed for the public generally. Bad land use everywhere, in all forms, stemmed from flaws in modern culture. Cultural change was needed. For that to come about, collective action was going to be needed. His particular hope late in his life was that every organization and profession in the many fields relating to land use—including agronomy, forestry, grazing, and water management— might join together in common messages that presented nature and the human plight in new ways. A united front, with common messages that were endlessly repeated, was the only pathway to change that Aldo Leopold could imagine.

From the beginning of his professional life Leopold was active in numerous organized efforts to labor for change. He was pleased to sign on to organized efforts with a variety of aims, all in the name of pushing for more sensible ways of dwelling in nature. In his personal life, Wendell Berry has also taken interest in organized conservation efforts, including public protests and lobbying. In his writing, however, Berry has downplayed such work, if not ignored it. Indeed, as noted, he has at times spoken harshly of movements of all stripes. The message in his writing is that reform needs to take place chiefly at the individual level, even as people work as neighbors on projects of common interest (such as community-supported agriculture arrangements). In his work, David Orr emphasizes the need for design and planning at the neighborhood scale. But he too—perhaps influenced by Wendell Berry's example—has had little to say about the need for an overall organized movement. No more than Berry has he pushed readers and audience members to organize and fight collectively, using new ways of thinking and valuing, to press for legal and policy reform.

The need for collective action could hardly be more manifest than it is in Hardin's tale of the commons. The only solution to this kind of tragedy is for grazers to get together and work in tandem. It is hard to conceive of any real solution that looks chiefly (or only) for grazers as individuals to become better people. (Hardin considered the option and promptly dismissed it.) The need for collective action in Hardin's story becomes even weightier when we consider, as Hardin did not, that the grazers can succeed in their rule-making work only if people outside the commons leave them free to do so—one of many points added by Elinor Ostrom and colleagues.[3] The grazers will need to be embedded within larger-scale governance regimes, ones that empower them to control their local lands and to exclude outsiders. As for the grazers acting responsibly, we need to consider the spillover effects of their grazing on lands outside their region. Good land use at the larger scale will require the grazers to pay attention to such external effects and to mitigate the harmful ones. Will they do so when there is no larger-scale governance arrangement—itself the product of well-guided collective action—that imposes such a duty on them?

Collective action is every bit as needed in the work of protecting wilderness areas and in redefining property rights. Wilderness areas by definition enjoy protection only when the liberties of all people are constrained, only when everyone is told to enter them only as recreational visitors. Legal restraints of this type can only come about when people work together and when civic-minded citizens, possessed of good visions, invite other people to share them. It is not enough, plainly, for wilderness advocates simply to talk up the many reasons why wild places should be protected. Mutual coercion is needed, in this setting just as much as in Hardin's tale. As Reverend Martin Luther King put it, love without power is mere sentiment. As for private property rights, they are themselves the products of social convention incorporated over time into laws. Take away the laws, as Jeremy Bentham famously said, and property rights disappear.[4] Without coercive power in some form (and varied forms are possible), property rights regimes simply do not work. And absent an autocracy, changes in existing property laws can only come about through orchestrated reform efforts. It is not sensible to hope that lawmakers on their own will awaken some day and, on their own initiative, enact revisions to current laws simply because the changes make good sense.

Hardin was wrong to imply (or, more aptly, his readers were wrong to infer) that a commons is a special kind of landscape, distinguishable from landscapes divided into privately owned parcels. All landscapes display traits of a commons, even when every square meter is privately owned. Water, air, wildlife, nutrient flows, fire and storm regimes, and much more unite landscapes in ways that ignore the surveyor's artificial lines. To the extent that they do, the lessons of Hardin's tale apply everywhere: whenever the human footprint threatens the land's capacities, mutual coercion in some form becomes needed.

Mutual coercion, as already seen, needs to emerge as the final step in a process of study and reflection that takes account of the best ecological science and that helps people draw upon their best ideas about good land use. In congested landscapes—which is to say in the modern era—good land use is an emergent property that arises at the community level, when and only when the community functions well.

The examples here—of property rights, wilderness, and the tragedy of the commons—are typical of environmental problems in that the work of correcting and averting errors calls for concerted effort. It needs remembering, though, that the collective effort most needed, the one that provides the base for all others, is the effort at promoting new ways of seeing and valuing nature—just the work that Aldo Leopold undertook with increasing focus in his late years and just the work Pope Francis has now so strongly encouraged. Through some means, somehow, new ways of seeing and valuing nature need to be put forward, and forcefully, so that they spread widely and sink deeper social roots. Without this kind of change, without large-scale pushes in this direction, it is hard to see how the local actions that Berry, Orr, and many others support will take place in any widespread way. Local projects, to be sure, can themselves be part of the cultural reform process, just as wilderness areas and wildlife refuges can be, when they are managed and justified with that aim in mind. But the foundational work—the identification of root causes, the formulation of new ways of seeing, valuing, and talking—seems highly dependent on the kind of orchestrated reform effort that the late Leopold envisioned.

This returns us to Wendell Berry and to the various theories of progress. Berry, as noted, fits within the moral worldview of responsible individualism that characterized Jacksonian democracy at its best. It was a world-view that supposed progress would unfold if individuals were simply free to pursue their own interests and acted within a moral frame inherited from Christianity. Berry, though, has shown little faith in the invisible hand of the market; he is not one to suppose that the market's special powers can transform what Christianity would have deemed vice—the vigorous pursuit of self-interest—into something like public virtue. When that mechanism of change is removed, it is then unclear how progress is supposed to come about. The inexorable, large-scale forces that Karl Marx thought would drive society onward—an understanding that Marx, in turn, got from Hegel—offer little promise of bringing good land-use results, not in anything like the time horizon in which reform needs to unfold. If such forces cannot be counted on, if the market's wonders are also not adequate, then that leaves only the possibility of people acting together in conscious ways. This means some form of what is sometimes termed civic republicanism—of Stoicism—of people somehow rising about their self-interests and working collectively for their shared good.

In the end, collective action is perhaps most needed because it is only when people work in concert that it becomes possible to talk meaningfully about our need to respect nature's limits, and for the reasons, again, that Hardin illustrates. It is sometimes said that we merely need to get people to change their ways of living, showing more ethical restraint as individuals. This wishful reasoning, though, is terribly flawed, and the mere fact that it would draw support is further evidence of our excessive commitment to competitive individualism. Nature's limits don't apply to individuals as such; they apply to humankind as a whole. It is humanity as such that needs to respect them, all the people living within a landscape. To expect all individuals as individuals to act wisely is the most starry-eyed of utopian visions and should be ridiculed as such. It is a steep enough moral challenge to get people to work with their neighbors, doing their fair share, to promote the welfare of all while demanding participation also by the recalcitrant. In that setting, costs and benefits are fully aligned. When individuals act alone costs and benefits split apart, free-riders charge ahead, many land-use options disappear, senses of community dissipate, and market forces reward the worst-behaving. In any event, we've tried this option for decades. It does not work.

What to Do

The aim of this inquiry has chiefly been to probe our environmental plight, intellectually and morally, by drawing wisdom from our most perceptive observers and by revisiting key enigmas. Yet, the many themes and observations that have recurred along the way do outline or at least imply a new, long-term reform effort, more likely to foster long-term flourishing than our current efforts.

Our misuses of nature, to reiterate, have much to do with the ways that we think and talk about who we are, what nature is, and how we relate to it and to each other. To the extent this is true—as Leopold, Berry, Orr, Pope Francis, and others have insisted—then we need in coming years and decades to think and talk in better ways and then, as best we can, to live consistently with that new orientation. Long-term reform, accordingly, should aim at promoting better ways to think and talk. It should aim above all at cultural change.

Current environmental reforms have largely stalled, due in important part to corrupted politics but due as much or more to the disconnection between modern culture and the kinds of nature-sensitive policies that green advocates propose. The path ahead, the path of hope, needs to reflect this clashing reality and respond to it.

For starters, environmental advocates themselves need to think and talk in new ways, in the ways outlined by visionaries profiled here, even if that means, for the time being, that their ideas and statements draw little assent from audiences, in more holistic ways, openly laced with values, openly presenting alternatives to short-sighted, competitive individualism and the institutions (private property, the market) shaped around it. Organizations will need to do the same as they go about presenting their values, hopes, and proposals. The opposing approach, the one employed far too often and long, is simply to use language that makes sense to ordinary people, where they now are—to appeal to them in ways that accept and strengthen their current worldview. But how sensible can that be if that worldview sustains and guides the bad behavior?

It hardly needs saying that journalists and writers in all settings could aid the effort by changing how they frame events and issues, by using new intellectual and moral foundations as the bases of their thinking and writing. They should do so, and environmental advocates should press them hard on it. To continue writing about environmental issues in familiar, worn ways is to keep us in our current rut, to side with the forces of degradation and to help frustrate reform efforts. It is nothing like neutral or responsible journalism.

It would hardly be wise for all environmental groups simply to stop doing what they are doing. But they need to realize that their separate efforts are very likely not now well aimed. They do not seek, first and foremost, to foster the kinds of culture change upon which long-term success depends.

What if the 100 largest conservation organizations got together to think critically about modern culture and to craft new, shared means of pushing forward new ways of thinking and talking?

What if they decided to turn away from their current approach—the fragmented, thousand-points-of-light mentality—and decided instead to forge their individually small lights into a beacon of far greater strength, just as Aldo Leopold recommended?

What if they funded think-tanks, and held conferences, to engage the root causes and to hone their ways of confronting them, shaping messages perhaps much like those put forth by Pope Francis?

The social and intellectual fragmentation that helps account for our misuses of nature shows up clearly today within the environmental movement itself, which is a classic liberal cause chiefly in that it is populated by groups and individuals guided by a go-it-alone mentality. On this point, Wendell Berry might well be right in instructing environmental organizations to judge themselves first and begin reform from within. Too much environmental rhetoric accepts and employs modes of thought that make bad practices seem more reasonable; rhetoric that endorses individualism, short-term horizons, instrumental rationality, market-based assessments of nature, moral value limited to humans, and the like. The motives involved are understandable; the need is there, it seems, to push policy proposals that might gain early acceptance and to talk to people in power in ways they know and understand. But how is real change then supposed to happen?

Audiences might be more receptive than environmental groups realize for fresh ways to think and talk. In time they might respond favorably, just as Pope Francis contends. The new messages, though, need to be heard again and again, and in ways that soon make them sound familiar. This good outcome, though, requires coordinated effort of a type rare in today's environmental movement. And it requires advocates to emphasize cultural issues, ways of thinking and talking—not facts, not technologies, not legal or economic tools—to an extent they never have.

Among the benefits of such an organized approach is that it targets something that we share: our dominant culture and ways of thinking. It puts the blame on the system as such, on all of us, not on particular bad actors who quite often simply strive to survive in a system that pushes them to act as they do. It is good also in that it can invite people to join in the conversation and to interact on the key issues, a possibility much facilitated by the Internet and electronic networks. Further, it can help the environmental cause regain the high moral ground while distinguishing its stances more clearly from those of industry and resource-production groups that view and treat nature in much different ways.

Too much green rhetoric calls individuals to make sacrificial changes in their personal lives, implicitly blaming them for ecological ills and ignoring the constraints that they face. A sounder approach is to challenge the systems in which we are all embedded, much as Pope Francis has done. The call to individuals should not be to buy green products or cut back a bit here and there. Instead it should be to engage the cultural issues and begin talking and thinking in new ways while inviting others to join them. Yes, individuals as such do need to make changes. But the changes are in the ways they think, talk, value, and hope, in the ways that see nature and understand their place in it, in the moral standards that they hold up for themselves and others, and in their willingness to join the cause for the resettlement of their natural homes.

A New Reform Trajectory

It needs knowing and saying, however, that these various cultural shifts do not at all line up well with the dominant social reform rhetoric of the present. Indeed, the clash is substantial and troubling. For the past 250 years, the dominant reform trajectory has been guided by the idea of individual rights, particularly negative individual liberty, enjoyed equally by all. Major reform efforts, intended to enhance rights and equality, have drawn upon and extended this reform trajectory, from the abolition of slavery (and even before, to broadened white male suffrage and enhanced property rights for married women) through the latest campaign for marriage equality. So successful has this reform rhetoric been over time that one is tempted to think it supplies a template for environmental change. The reality, though, is much to the contrary.

The civil rights, equality line of thinking is highly if not fully anthropocentric and firmly grounded in human exceptionalism. It holds high the morally worthy individual as such, and even reform efforts that address interpersonal ties—gay marriage the most recent—commonly charge ahead stressing individual autonomy. The language of choice, the language that flows almost inevitably, is strong on individual rights and weak or silent on duties. The community as such makes little appearance in it. Similarly the common good essentially disappears, displaced (as it is in market reasoning) by the summed preferences of individuals as such. Normative choices are put within the sphere of the individual, not treated as public business for public resolution. It goes without saying that, in this reform trajectory, the entire natural world makes essentially no appearance. The relatively rare exceptions have to do with individual claims to greater shares of nature, to access to land or clean water, so that the claimants too might participate more equally in the destructive system. Interconnections and interdependencies of the type promoted by Berry and Pope Francis hardly carry much weight. As in the market, the focus is on the present with little concern for the future.

What this means, alas, is that, for environmental reformers attuned to the root cultural problems, the long-continuing civil rights crusade is not a kindred cause. True kindred spirits would speak the much-different language of community and duty. They would talk about humans as embedded creatures. They would hold high the notion of the common good, even as they might debate its content. The rhetoric of individual rights, of liberty and equality, hardly ought to be set aside; it has done much good and can do even more. But it needs to be kept in its rightful place. And that place needs to remain rather separate from issues of humans in nature, separate from the ways we think and talk about ourselves as fellow members of the land community. It needs to remain largely silent as the call goes out, as it must, to curtail individual liberties (property rights quite often) when required to keep our common home healthy, biologically diverse, and beautiful.

To get to this point is to see that the kinds of cultural changes now needed do not simply entail the expansion and dominance of today's progressive reform thinking. Far from it. The cultural elements that require change are more pervasive than that. Indeed, many ardent civil-liberties advocates will be among the citizens whose worldviews are most overdue for revision. The far political left as much as the far right is home to strident calls for liberty and opportunity, particularly for the poor, which treat a healthy environment (if honored at all) as a secondary concern for some later era to address. Environmental ills with disparate racial impacts are treated primarily if not solely as issues of race discrimination, again discounting or ignoring the effects on the natural home of everyone (and sometimes fostering social divisions in the process). The far left can distrust and undercut shared governance almost as well as the far right; it too can challenge and reject holistic moral visions, calls to community membership, and perhaps most distressing, the whole idea of expecting people to act virtuously. No combination of negative liberty and equality, no celebration of cultural diversity, provides much help for the journey that lies ahead.

There is much truth, then, in the claim by Pope Francis (*Laudato Si'*, ¶ 202) that "a great cultural, spiritual and educational challenge stands before us, and it will demand that we set out on the long path of renewal." By "us" we need to read "all of us," not just industry advocates and political conservatives. We are similarly wise to take literally and seriously Francis's summation (¶ 194) that the path ahead, "put simply . . . is a matter of redefining our notion of progress." The progress he challenges is not just the supposed progress mismeasured through economic calculations of gross domestic production. It is also the progress now measured in terms of enhanced individual liberty to act as one wants, even if it means degrading nature, disrespecting moral values in the world, and withholding gratitude for the natural gifts we share and tend. It is progress that seeks to expand opportunities so that all individuals, regardless of personal traits, have fair chances to share in the overconsumption. And it is progress that pushes out of the public arena the inherited wisdom of the ages, particularly when embedded in the language of religion.

The call of Pope Francis, the call of other leading voices, is that we need to stop dwelling on the surface. We need to probe beneath the outward signs of decline, the superficial causes, to plumb the depths. Francis believes, we too must soon believe, that we have the learning and reasoning abilities to understand where we are—to escape the tyranny of the present, in Cicero's remembered words—and to identify and address our flaws.

We are not bad people at the core. We are, however, people who live, breathe, and make sense of the world immersed in the culture of our times; the water in which we swim. Using its language and tropes we navigate our ways, interpreting what we view and judging the wise and foolish. Particularly in the United States, the prevailing culture reflects still the reality of colonial days, when a small number of people encountered a vast continent of seemingly endless wealth.[5] Fossil fuels had not been harnessed. The steam engine was yet to come. The chemical age had not dawned. The evil of the day was the exploitive social hierarchy, political and social, and mass culture change unfolded in response to that evil.

Given the vast lands and resources, given the comparatively simple technology, the sensible course for the dissatisfied of that much-different era was to promote liberty and equality among all people, leaving them free to seek their ways and live as they saw fit. If they achieved that goal, the hierarchies would become less oppressive and some of them would tumble. A secondary language did live on, to be sure, in the language of family, community, faith, and obligation. But it was pushed down, and for reasons and under circumstances that seemed sufficient.[6]

We can be grateful to our Revolutionary era forebears and their successors who reshaped the then-dominant culture in ways that have brought much good. But we pay them tribute best not by bowing down to the cultural principles useful and sensible in their vastly different time. We do so by following their revolutionary example and rising up ourselves to challenge the flawed culture of our time. Our culture now is in need of major revision every bit as much as were the quasi-feudal, oligarchic hierarchies of the early modern era. Without losing the great gains of our recognition of individual moral worth, without losing sight of the polestar of equality, we nonetheless need to revise our worldview to account for our crowded, overused world and for our vastly greater technical powers to degrade it.

A worldview sound for our time would put far greater stress on our interconnections and interdependencies. It would keep the language of the common good in daily use. It would see all uses of nature in all places as matters of public concern. It would see the recurring need for collective action to avert the full array of looming land-use tragedies. It would reattach individual rights to the common good and redefine them so that they promoted that good, even as they treated individuals with dignity. Our "second language" of community, responsibility, and cooperation would become, in our dealings with nature, our primary language. And our community would include other life forms, natural processes, and human generations of the past and the future.

The good use of nature is conceivable within such a culture. It is not conceivable within the one we now have.

ACKNOWLEDGMENTS

My work on this book reaches back to my first years in the academy, now a full three decades ago, and it is hardly possible to acknowledge the many debts I have accumulated along the way, particularly to friends and colleagues who have helped me move along. I feel particularly grateful to Clark Bullard, J. Baird Callicott, Jonathan Cobb, Bruce Hannon, Heidi Hurd, Robert McKim, Curt Meine, Joseph Sax, J. Michael Scoville, Julianne Warren, Amy Wildermuth, Todd Wildermuth, and Donald Worster. Versions of several chapters have previously appeared in print and I thank the editors of the various publications for assisting me with them and the conference organizers for inviting me to speak. Chapter 1 appeared in substantially similar form in the *Washington Journal of Environmental Law and Policy* in 2012. Chapter 2 appeared in somewhat different form and with a different title in Jason Peter, ed., *Wendell Berry: Life and Work* (Lexington: University Press of Kentucky Press, 2007). Chapter 5 appeared in David Grinlinton and Prue Taylor, eds., *Property Rights and Sustainability: The Evolution of Property Rights to Meet Ecological Challenges* (Leiden: Martinus Nijhoff, 2011). Chapter 6 was presented at a conference celebrating the fiftieth anniversary of the Wilderness Act of 1964, and was published in the law school's journal, *Environmental Law*, in 2014. Chapter 7 was presented at a conference on the tragedy of the commons at Brigham Young University in 2014, and appeared in the *Brigham Young University Law Review* in 2015.

It is with considerable gratitude that I dedicate the book to the many good people who have supported the Illinois-based Prairie Rivers Network for the nearly fifty years of its existence, enabling it to provide a strong voice for the waterways, people, and wildlife of Illinois and to foster cultural change.

NOTES

CHAPTER ONE

1. Leopold's life is recounted most ably in Curt Meine, *Aldo Leopold: Life and Work* (Madison: University of Wisconsin Press, 1988). The fullest single treatment of Leopold's evolving conservation thought, covering his scientific understandings, philosophic groundings, and cultural criticism, is Julianne Lutz Newton, *Aldo Leopold's Odyssey* (Washington, DC: Island Press, 2006).
2. Aldo Leopold, "Engineering and Conservation" (1938), in *The River of the Mother of God and Other Essays by Aldo Leopold*, ed. Susan L. Flader and J. Baird Callicott (Madison: University of Wisconsin Press, 1991), 254. This volume contains important writings by Leopold, some published in journals during his time, many published here for the first time. New writings by Leopold also were put out in Leopold, *For the Health of the Land: Previously Unpublished Essays and Other Writings*, ed. J. Baird Callicott and Eric T. Freyfogle (Washington, DC: Island Press, 1999).
3. A dozen or more of Leopold's late talks were published, either during his lifetime or later. They are contained in the sources mentioned in n. 2. Unpublished talks and the manuscripts of published ones are found in the Leopold Archives at the University of Wisconsin, Madison, which has organized Leopold's papers in an archival series identified with the prefix 9/25/10. They are online at http://uwdc.library.wisc.edu. The many boxes in that series are divided into thirteen categories, by type of document. Thus, Leopold's "writings" are in the group numbered 10-6, in the sequence 10-1 to 10-13. Each group is divided into boxes, and boxes into folders. The online index goes further, designating items in each folder by item number. The online lists of items, however, can confuse because the items in a folder often do not appear in the order listed and the lists are not always complete. Typically, however, all items in a folder are numbered consecutively, so it is possible to locate an item using the box number, folder number, and page number. These page numbers do not appear on the documents in their hard-copy form in the archives. Instead, they are generated by the online display of the documents in digital form. Thus, the page numbers cited here are useful in quickly locating a document in the online archive, but a researcher undertaking a hand search for an item in the archives would need to search by hand through all of the items in a particular file.

    The citation format used here identifies each item by group, box, file folder, and page number, using the computer-generated page number within the folder. (Often, pagination runs consecutively among multiple folders in a given box.) As an example: The archives contain the outline of a talk that Leopold delivered to the Friends of the Native Landscape on March 26, 1946. It is found at 9/25/10-6: Writings, box 14, folder 2, page 122. Citations below follow an abbreviated format (using the same example): ALA (for Aldo Leopold Archives) 10-6, box 14, folder 2, 122. Manuscripts and note cards of Leopold's talks are found throughout the Leopold Archives. Many of the later lectures appear in box 14, folders 2 and 3. An incomplete list of lectures, all but one from 1935 or later, is at ALA box 14, folder 3, 419–20. This list of some eighty-five lectures excludes not just earlier radio and extension talks but lectures chiefly prepared for classroom delivery; many of the latter are in box 15, folders 3 and 6, and a few were used in this assessment. It was challenging for the archive organizers to distinguish between lecture manuscripts and other writings loosely termed "unpublished writings." The latter, which contain many lecture-related items, are in box 14; in the case of handwritten items, typed versions are often found in box 17 and/or 18. For the most part, items designated as unpublished manuscripts rather than lectures—even when a notation on the manuscript indicates that a manuscript was used for a talk—are not included in the list in box 14, folder 3.

4. Published posthumously by Oxford University Press in 1949.
5. Leopold, "Planning for Wildlife," in Leopold, *For the Health of the Land*, 198.
6. Leopold, "The State of the Profession," in *River of the Mother of God*, ed. Flader and Callicott, 280.
7. E.g., Leopold, "Land-Use and Democracy," in ibid., 298; "Armament for Conservation," ALA 10-6, box 16, folder 6, 692.
8. Leopold, "The Meaning of Conservation," ALA 10-6, box 17, folder 7, 1293 (note cards prepared for a talk that was likely given more than once).
9. Leopold, "Conservationist in Mexico" (3×5 notecards), ALA 10-6, box 14, folder 3, 470.
10. "The Farmer as a Conservationist," in *River of the Mother of God*, ed. Flader and Callicott, 257–58 (first delivered to a "Farm and Home Week" audience).
11. Leopold, "Biotic Land Use" (3×5 note cards), ALA 10-6, box 14, folder 3, 451. A full text version of this talk, one of Leopold's most important discussions of land health, has appeared in *For the Health of the Land*, 198.
12. Aldo Leopold, *A Sand County Almanac and Sketches Here and There* (New York: Oxford University Press, 1949), viii, 204.
13. Leopold, "The Role of Wildlife in a Liberal Education," in *River of the Mother of God*, ed. Flader and Callicott, 303.
14. Leopold, "On a Monument to the Pigeon," ALA 10-6, box 9, folder 7, 762 (from Leopold edit dated August 25, 1946) (delivered to a birding group).
15. Leopold, "The Basis of Conservation Education," ALA 10-6, box 17, folder 5, 999 (first given to a Kiwanis Club gathering in 1939).
16. Leopold, "State of the Profession," 280.
17. Leopold, "Game Policy Model 1930," ALA 10-6, box 14, folder 2, 318.
18. Leopold, "Conservation: In Whole or in Part?," in *River of the Mother of God*, ed. Flader and Callicott, 311.
19. Leopold, "Land Pathology," in ibid., 217.
20. Leopold, "Conservation: In Whole or In Part?," 310.
21. On Leopold plans for this talk, see Julianne Lutz Warren, "Science, Recreation, and Leopold's Quest for a Durable Scale," in *The Wilderness Debate Rages On: Continuing the Great New Wilderness Debate*, ed. Michael P. Nelson and J. Baird Callicott (Athens: University of Georgia Press, 2008), 97.

22. Leopold, "The Land-Health Concept and Conservation," in *For the Health of the Land*, 219. On Leopold's forthcoming address, see Warren, *Aldo Leopold's Odyssey*, 350.
23. Leopold, "Land-Health in S.W. Wisconsin," ALA 10-6, box 14, folder 2, 202 (delivered to a civil engineering gathering in November 1943).
24. Leopold, "The Ecological Conscience," in *River of the Mother of God*, ed. Flader and Callicott, 342.
25. Leopold, "Basis of Conservation Education," ALA 10-6, box 16, folder 5, 549.
26. Leopold, "Biotic Theories and Conservation," ALA 10-6, box 14, folder 2, 301.
27. Leopold, "Conservationist in Mexico" (3×5 notecards), 471 (May 1941).
28. Leopold, "Biotic Land-Use," in *For the Health of the Land*, 202.
29. Leopold, "Role of Wildlife in a Liberal Education," 303.
30. Leopold, "Conservation: In Whole or in Part?," 317.
31. Leopold, "Meaning of Conservation," 1293.
32. Leopold's role is explained in Meine, *Aldo Leopold*, 480–81.
33. Leopold, "Conservation," ALA 10-1, box 1, folder 14, 510.
34. Leopold, "Conservation: In Whole or in Part?," 311.
35. Leopold, "The Role of Wildlife in Education," ALA 10-6, box 17, folder 7, 1313.
36. Leopold, "Foreword," ALA 10-6, box 17, folder 5, 1203 (from the original, longer foreword that Leopold wrote for his Almanac in July 1947 but then discarded in favor of the final, shorter one). This version was later published in J. Baird Callicott, ed., *Companion to "A Sand County Almanac": Interpretive and Critical Essays* (Madison: University of Wisconsin Press, 1987), 281.
37. Leopold, "Modus Vivendi for Conservationists," ALA 10-6, box 17, folder 6, 1127.
38. Leopold, letter to William Vogt, ALA 10-2, box 4, folder 11, 911.
39. Leopold, "Land-Use and Democracy," 300; Leopold, "Basis of Conservation Education," ALA 10-6, box 17, folder 5, 999.
40. Leopold, "Ecological Haves and Have-Nots," ALA 10-6, box 17, folder 6, 1108.
41. Leopold, "Meaning of Conservation," 1296.
42. Leopold, "Conservation: In Whole or in Part?," 318.
43. Leopold, "Ecological Conscience," 346.
44. Ibid., 338.
45. Richard Tarnas, *The Passion of the Western Mind: Understanding the Ideas That Have Shaped Our World* (New York: Ballantine Books, 1991), 441.
46. Leopold, "Ecology and Politics," in *River of the Mother of God*, ed. Flader and Callicott, 281.
47. Louis Menand, *The Metaphysical Club: A Story of Ideas in America* (New York: Farrar, Straus and Giroux, 2001), 236–37, 238–45.
48. Leopold, *Sand County Almanac*, 204, 205.
49. John Stuart Mill, *On Liberty* (1859; Indianapolis: Bobbs-Merrill, 1956), 68, 91–92, 95–100.
50. Tarnas, *Passion of the Western Mind*, 345–46, 417–18.
51. Leopold, *Sand County Almanac*, 205.
52. Leopold, "Conservation: In Whole or in Part?," 315.
53. Leopold, "Planning for Wildlife," 194.
54. Leopold, "A Biotic View of Land," in *River of the Mother of God*, ed. Flader and Callicott, 267.
55. Leopold, "Land-Health Concept and Conservation," 226.
56. Leopold, *Sand County Almanac*, 138.
57. Leopold, "State of the Profession," 277.

58. Luc Ferry, *A Brief History of Thought* (New York: HarperCollins, 2010), 20.
59. Quoted in Tarnas, *Passion of the Western Mind*, 44.
60. Ibid., 366–67.
61. See Warren, *Aldo Leopold's Odyssey*, 78–79; Meine, *Aldo Leopold*, 214–15.
62. Franklin L. Baumer, *Modern European Thought: Continuity and Change in Ideas, 1600–1950* (New York: Macmillan, 1977), 375–76.
63. Menand, *Metaphysical Club*, 200.
64. William James, *Pragmatism: A New Way for Some Old Ways of Thinking* (New York: Longmans, Green and Co., 1907), 76.
65. Menand, *Metaphysical Club*, 322. Pragmatism's influences on Leopold are mentioned in Bryan G. Norton, *Sustainability: A Philosophy of Adaptive Ecosystem Management* (Chicago: University of Chicago Press, 2005), 97. Norton's claim that Leopold was influenced by A. T. Hadley, a conservation disciple of William James, is sharply challenged in J. Baird Callicott et al., "Was Aldo Leopold a Pragmatist? Rescuing Leopold from the Imagination of Bryan Norton," *Environmental Values* 18 (2009): 453.
66. Leopold, *Sand County Almanac*, 221, 224–25.
67. Warren, *Aldo Leopold's Odyssey*, 336–43, 346–50.
68. Leopold, *Sand County Almanac*, viii; Richard Norman, *The Moral Philosophers* (New York: Oxford University Press, 1983), 71–93.
69. John Stuart Mill, *Utilitarianism* (1861; Barnes and Noble ed., 2005), 8, 12.
70. Mill, *Utilitarianism*, 55–57.
71. Ferry, *Brief History of Thought*, 24, 28.
72. Ibid., 31 (emphasis omitted).
73. Leopold, "Farmer as a Conservationist," 265.
74. Mill, *Utilitarianism*, 17, 19; Peter Singer, *Karl Marx* (New York: Oxford University Press, 2000), 80–84.
75. Terry Eagleton, *Why Marx Was Right* (New Haven, CT: Yale University Press, 2011), 86.
76. A prominent example was Clive S. Lewis, *The Abolition of Man*, 1st American ed. (New York: Macmillan, 1947).
77. Leopold, "Ecology and Politics," 281.
78. Leopold, "The Statesmanship of E. Sydney Stephens," ALA 10-6, box 16, folder 8, 1227.
79. Leopold, "Conservation: In Whole or in Part?," 317.
80. Leopold, "Conservation and Politics," ALA 10-6, box 16, folder 6, 633.
81. Leopold, "Role of Wildlife in a Liberal Education," 303.
82. Leopold, "Conservation Education: A Revolution in Philosophy," ALA 10-6, box 17, folder 6, 1107. On the dating, see Susan Flader, *Thinking Like a Mountain: Aldo Leopold and the Evolution of an Ecological Attitude Toward Deer, Wolves, and Forests* (Columbia: University of Missouri Press, 1974), 206.
83. Leopold, "Ecological Conscience," 342–43.
84. Leopold, "Armament for Conservation," ALA 10-6, box 16, folder 6, 691. Leopold omitted the word when he arranged his ideas for an article appearing in *Audubon Magazine*. Leopold, "Land-Use and Democracy," 295.
85. Leopold to Walter John, ALA 10-6, box 16, folder 6, 708.
86. Leopold, "Game Policy Model 1930," 316.
87. Leopold, "Land Pathology," 214.
88. Leopold, "Motives for Conservation," ALA 10-6, box 14, folder 3, 43.
89. Leopold, "Conservation," 510.
90. John Bellamy Foster, *Ecology against Capitalism* (New York: Monthly Review Press, 2002), 59.

91. Leopold, "Land Pathology," 215.
92. Leopold, "Role of Wildlife in Education," 1315.
93. Leopold to William Vogt, ALA 10-2, box 4, folder 11, 911.
94. Leopold, "Farmer as a Conservationist," 259.
95. Leopold, "Wherefore Wildlife Ecology?," in *River of the Mother of God*, ed. Flader and Callicott, 337.
96. Leopold, "Engineering and Conservation," 254.
97. Leopold, "Armament for Conservation," 693.
98. Tarnas, *Passion of the Western Mind*, 34.
99. Singer, *Karl Marx*, 81.
100. Eagleton, *Why Marx Was Right*, 68.

CHAPTER TWO

1. My thinking about Berry's work has been much informed by the particularly thoughtful study by Kimberly K. Smith, *Wendell Berry and the Agrarian Tradition: A Common Grace* (Lawrence: University Press of Kansas, 2003), and also by Fritz Oehlschlaeger, *The Achievement of Wendell Berry: The Hard History of Love* (Lexington: University Press of Kentucky, 2011). Early good studies, more focused on his literary techniques, include Andrew J. Angyal, *Wendell Berry* (New York: Twayne Publishers, 1995), and Janet Goodrich, *The Unforeseen Self in the Works of Wendell Berry* (Columbia: University of Missouri Press, 2001). Also useful is a volume edited by Jason Peters, *Wendell Berry: Life and Work* (Lexington: University Press of Kentucky, 2007), although few contributors to the volume (in which an earlier version of this chapter appeared) challenged Berry's thought in any significant way.
2. Wendell Berry, *Recollected Essays* (San Francisco: North Point Press, 1981), ix.
3. Wendell Berry, *What Are People For?* (San Francisco: North Point Press, 1990), 149, quoting Albert Howard, *The Soil and Health: A Study of Organic Agriculture* (New York: Devin-Adair, 1947), 11.
4. Wendell Berry, *A Continuous Harmony: Essays Cultural and Agricultural* (New York: Harcourt Brace Jovanovich, 1972), 164.
5. Wendell Berry, "Health Is Membership," in Berry, *Another Turn of the Crank* (Berkeley, CA: Counterpoint Press, 1995), 86, 90.
6. Wendell Berry, "Contempt for Small Places," in Berry, in *The Way of Ignorance* (Washington, DC: Shoemaker and Hoard, 2005), 7.
7. A perceptive meditation on Adam Smith and his inattention to nature is offered in the title essay of Donald Worster, *The Wealth of Nature: Environmental History and the Ecological Imagination* (New York: Oxford University Press, 1993), 203–19.
8. Marxism, it should be clear, stands quite far from the version of communism embodied in the former Soviet Union (which was in essence a form of state bureaucratic capitalism having little to do with the writings of Marx). In several writings, Berry makes clear that his critique of industrial capitalism applies with essentially the same force to the brutal industrialism of state-run economic systems, e.g., "The Total Economy," in *Citizenship Papers* (Washington, DC: Counterpoint, 2003). Marx held quite a low opinion of farm workers and small-farm owners—another contrast with Berry—and had little hope that they would play a significant role in fostering the dominance of labor over capitalist owners.
9. Karl Marx and Frederick Engels, *The Communist Manifesto*, ed. Eric Hobsbawm (London: Verso, 1998), 37.
10. Genovese's writings on the subject include *The Southern Tradition: The Achievements*

and Limitations of an American Conservatism (Cambridge, MA: Harvard University Press, 1994).

11. For instance, Burley's important comment in "The Wild Birds": "The way we are, we are members of each other. All of us. Everything. The difference ain't in who is a member and who is not, but in who knows it and who don't." Wendell Berry, *The Wild Birds: Six Stories of the Port William Membership* (San Francisco: North Point Press, 1986), 136–37.

12. Michael Sandel, *Public Philosophy: Essays on Morality and Politics* (Cambridge, MA: Harvard University Press, 2005), 10.

13. Ibid., 20.

14. Berry, "Way of Ignorance," in *Way of Ignorance,* 62.

15. Wendell Berry, "Out of Your Car, Off Your Horse," in Berry, *Sex, Economy, Freedom and Community* (New York: Pantheon, 1993), 24.

16. Berry, "Way of Ignorance," 63.

17. Maury Klein, *The Flowering of the Third America: The Making of an Organizational Society, 1850–1920* (Chicago: Ivan R. Dee, 1993), 104.

18. Wendell Berry, "The Purpose of a Coherent Community," in Berry, *Way of Ignorance,* 69.

19. I explore the tragedy in *The Land We Share: Private Property and the Common Good* (Washington, DC: Island Press, 2003), 157–78.

20. For instance, Elinor Ostrom, *Governing the Commons: The Evolution of Institutions for Collective Action* (Cambridge: Cambridge University Press, 1990).

21. Wendell Berry, "The Pleasures of Eating," in Berry, *What Are People For?,* 145.

22. Wendell Berry, "Conserving Forest Communities," in Berry, *Another Turn of the Crank,* 25.

23. The point is made in Robert N. Bellah et al., *Habits of the Heart: Individualism and Commitment in American Life,* updated ed. (Berkeley: University of California Press, 1996), xxiii.

24. Smith, *Wendell Berry and the Agrarian Tradition,* 71, 86.

25. Ibid., 117–20.

26. Berry, "Way of Ignorance," 67.

27. Wendell Berry, "Conservation and Local Economy" in Berry, *Sex, Economy, Freedom and Community,* 12.

CHAPTER THREE

1. All citations in this chapter are from David Orr, *Hope Is an Imperative: The Essential David Orr* (Washington, DC: Island Press, 2010).

CHAPTER FOUR

1. For instance, Worster, *The Wealth of Nature;* J. R. McNeill, *Something New Under the Sun: An Environmental History of the Twentieth-Century World* (New York: W. W. Norton 2000).

CHAPTER FIVE

1. The writing on private property is quite considerable, although little of it pays close attention to nature and our needs to live on land without abusing it. Since the middle of the twentieth century too little of it also has shown adequate awareness of the moral complexity of the institution and the link, once well known, between individual rights and the common good. Indeed, writers in recent decades have typically taken private property for granted, with no need to justify it, and used particular

visions of strong private rights as tools to curtail governmental powers. On the need to justify private property, and the various lines of argument used to do so, a still-valuable inquiry is Lawrence C. Becker, *Property Rights: Philosophic Foundations* (London: Routledge and Kegan Paul, 1977). The story of American ideas about private property, and their evolution over time, is usefully started with Gregory Alexander, *Commodity and Propriety: Competing Visions of Property in American Legal Thought, 1776–1970* (Chicago: University of Chicago Press, 1997); Stuart Banner, *American Property: A History of How, Why, and What We Own* (Cambridge, MA: Harvard University Press, 2011) (a sound work but with little coverage of owning nature); and William B. Scott, *In Pursuit of Happiness: American Conceptions of Property from the Seventeenth to the Twentieth Century* (Bloomington: Indiana University Press, 1977).

2. Richard Schlatter, *Private Property: The History of an Idea* (London: George Allen and Unwin, 1951), 265–69.

3. Ibid., 245–50.

4. Eric T. Freyfogle, *On Private Property: Finding Common Ground on the Ownership of Land* (Boston: Beacon Press, 2007), 84–87.

5. Freyfogle, "Property and Liberty," *Harvard Environmental Law Review* 34, no. 1 (2010): 107–14.

6. Freyfogle, *The Land We Share*, 37–85; Alexander, *Commodity and Propriety*, 91–240; Scott, *In Pursuit of Happiness*, 53–135.

7. Freyfogle, *On Private Property*, 29–57.

8. Jerry L. Anderson, "Britain's Right to Roam: Redefining the Landowner's Bundle of Sticks," *Georgetown International Environmental Law Review* 19 (2007): 375; John A. Lovett, "Progressive Property in Action: The Land Reform (Scotland) Act 2003," *Nebraska Law Review* 89 (2011): 739. A broader comparative survey is offered in Jerry L. Anderson, "Comparative Perspectives on Property Rights: The Right to Exclude," *Journal of Legal Education* 56 (2006): 539.

9. Freyfogle, *On Private Property*, 87–91.

10. Alexander, *Commodity and Propriety*, 21–88.

11. J. B. Ruhl, Steven Kraft, and Christopher Lant, *The Law and Policy of Ecosystem Services* (Washington, DC: Island Press, 2007).

12. Freyfogle, *On Private Property*, 91–104.

13. Freyfogle, *The Land We Share*, 203–53.

14. Mary Christina Wood, *Nature's Trust: Environmental Law for a New Ecological Age* (Cambridge: Cambridge University Press, 2014), presents a powerful argument for an expanded use of the public trust doctrine.

15. A classic ruling of the California Supreme Court, refusing to protect a water use deemed unreasonable, is Joslin v. Marin Municipal Water District, 429 P.2d 889 (1967). An important ruling sustaining the regulatory powers of the state to give meaning to reasonableness is Imperial Irrigation District v. State Water Resources Control Board, 225 Cal. App.3d 548, 275 Cal. Rptr. 250 (1990).

16. Freyfogle, *On Private Property*, 61–79.

17. Ibid., 105–30.

CHAPTER SIX

1. A rich collection of perspectives, debating whether wilderness has real existence or whether it is a social construction (and much more), is J. Baird Callicott and Michael P. Nelson, eds., *The Great New Wilderness Debate* (Athens: University of Georgia Press, 1998); Nelson and Callicott, eds., *The Wilderness Debate Rages On*.

2. The issue was given visibility in William Cronon, "The Trouble with Wilderness; or, Getting Back to the Wrong Nature," in *Uncommon Ground: Toward Reinventing Nature* (New York: W. W. Norton, 1995), 69, 81–83.

3. I develop my criticisms in "Back to Sustainability," in *Why Conservation Is Failing and How It Can Regain Ground* (New Haven, CT: Yale University Press, 2006), 113, drawing upon environmental historian Donald Worster's thoughtful comments in "The Shaky Ground of Sustainable Development," in Worster, *Wealth of Nature*, 142. An unusually solid version of sustainability appears in J. Baird Callicott, "Ecological Sustainability as a Conservation Concept," in *Beyond the Land Ethic: More Essays in Environmental Philosophy* (Albany: SUNY Press, 1999), 365, 366 (proposing that we sustain, in working landscapes, a version of Aldo Leopold's normative goal of land health).

4. A classic critique of "Enlightenment" as a concept and its associated liberal principles is Max Horkheimer and Theodor W. Adorno, *Dialectic of Enlightenment*, trans. John Cumming (1944; New York: Continuum Publishing, 1998). For a spirited conservative critique of liberalism, see John Gray, *Enlightenment's Wake: Politics and Culture at the Close of the Modern Age* (London: Routledge, 1995).

5. Jefferson's well-known answer in the Declaration of Independence was that the rights of men came from "their Creator." The Declaration of Independence para. 1 (U.S. 1776).

6. A longstanding practice of those opposed to environmental protections has been to push environmental values into the personal sphere, at times even comparing it to a religious choice manifestly ill-suited to ground public policy. An example is Andrew P. Morriss and Benjamin D. Cramer, "Disestablishing Environmentalism," *Environmental Law* 39 (2009): 309, 312–13. In a much-noted incident, then–Vice President Cheney made use of the tactic when proposing an oil-dependent national energy policy, ignoring options for energy conservation. Conservation, Cheney asserted, was properly a "personal virtue," not the basis for public policy. Joseph Kahn, "Cheney Promotes Increasing Supply as Energy Policy," *New York Times*, May 1, 2001, A1, A20.

7. See, for instance, Jonathan Haidt, "The Emotional Dog and Its Rational Tail: A Social Intuitionist Approach to Moral Judgment," *Psychology Review* 108 (2001): 814.

8. For instance, Edward O. Wilson, *The Social Conquest of Earth* (New York: W. W. Norton, 2012) (discussing the interplay between social characteristics and evolution in the formation of the human condition).

9. Norman, *The Moral Philosophers*, 71–93.

10. See, for instance, J. Baird Callicott, *In Defense of the Land Ethic: Essays in Environmental Philosophy* (Albany: SUNY Press, 1999), 17–27.

11. A classic early expression was Joseph Wood Krutch, *The Modern Temper: A Study and a Confession* (1929; New York, Harvest Press, 1956). Mary Midgley, *Can't We Make Moral Judgments?* (New York: St. Martin's Press, 1991), offers a useful survey and critique of the decline of public moral talk. Observers more inclined to perceive objective value embedded in nature include Robert Nozick, *Anarchy, State, and Utopia* (New York: Basic Books, 1974), and Iris Murdoch, *The Sovereignty of Good* (London: Routledge and Kegan Paul, 1970). Alasdair MacIntyre, *After Virtue*, 2nd ed. (Notre Dame, IN: University of Notre Dame Press, 1984), offers a more philosophically nuanced dissent in the natural law tradition.

12. Joseph R. DesJardins, *Environmental Ethics: An Introduction to Environmental Philosophy* (Belmont, CA: Wadsworth, 1993), 123–26.

13. For an illustrative critical response, see generally Richard A. Posner, "Animal Rights,"

Yale Law Journal 110 (2000): 527 (reviewing Steven M. Wise, *Rattling the Cage: Toward Legal Rights for Animals* [Cambridge, MA: Perseus Books, 2000]) (concluding that the argument of seeking to extend the moral worth of humans is logically inconclusive, and urging that animal welfare be enhanced instead by means of expanded property rights in animals—that is, greater commoditization—and a broader ban on cruelty to animals).

14. Examples include Tom Regan, *The Case for Animal Rights*, 2nd ed. (Berkeley: University of California Press, 2004), 329 (proposing that certain animals have rights); Peter Singer, *Animal Liberation: The Definitive Classic of the Animal Movement* (New York: Harper, 2009), 5–7 (proposing that utilitarian calculations include consideration of the utility of certain animals); Peter Singer, *The Expanding Circle: Ethics, Evolution, and Moral Progress*, rev. ed. (Princeton, NJ: Princeton University Press, 2011), 120–21 (proposing a fuller, influential argument for the continued extension of the category of morally worthy creatures).

15. Singer, *Expanding Circle*, 120.

16. Singer, *Animal Liberation*, 7–8.

17. John Christman, "Autonomy in Moral and Political Philosophy," *Stanford Encyclopedia of Philosophy*, http://plato.stanford.edu/entries/autonomy-moral/#IdeConSel.

18. Gina Kolata, "In Good Health? Thank Your 100 Trillion Bacteria," *New York Times*, June 14, 2012, A0.

19. J. Baird Callicott, "Ecology and Moral Ontology," in *The Structural Links between Ecology, Evolution and Ethics: The Virtuous Epistemic Circle*, ed. Donato Bergandi, Boston Studies in the Philosophy of Science 296 (Dordrecht: Springer Netherlands, 2013), 101 (noting similarities with earlier process-philosophy and process-theology approaches). See Scott F. Gilbert et al., "A Symbiotic View of Life: We Have Never Been Individuals," *Quarterly Review of Biology* 87 (2012): 325, 326–27.

20. Two useful surveys are Roderick Frazier Nash, *Wilderness and the American Mind*, 4th ed. (New Haven, CT: Yale University Press, 2001); Max Oelschlaeger, *The Idea of Wilderness: From Prehistory to the Age of Ecology* (New Haven, CT: Yale University Press, 1991).

21. See Hicham-Stéphane Afeissa, "Intrinsic and Instrumental Value," in *Encyclopedia of Environmental Ethics and Philosophy*, vol. 1, ed. J. Baird Callicott and Robert Frodeman (New York: Oxford University Press, 2009), 528, 530 (arguing that something has value "only if it is valued by a conscious being capable of intentionality").

22. Even those few who contend that some value adheres in nature and has real existence apart from the creative efforts of any sentient being are likely to admit that a sentient being is needed to recognize the value. For the purposes here, the labor of creating value and the similar labor of recognizing value that independently exists can be lumped together. Both require human involvement and decision making. See Affeissa, "Intrinsic and Instrumental Value," 530 (explaining that humans have the capacity to assign value "not only to themselves but also to other entities").

23. See Thomas Hurka, "Intrinsic Value," in *Encyclopedia of Philosophy*, 2nd ed., ed. Donald M. Borchert (New York: Oxford University Press, 2006), 719.

24. The Clean Air Act misuses science when it instructs EPA to issue ambient air quality standards to protect human health based solely on science. As commonly understood, the statutory provision excludes factors based on economics. But the governing standard, "requisite to protect the public health" with "an adequate margin of safety," is one that can be clarified and applied only through normative reasoning. 42 U.S.C. § 7409(b)(1) (2012). When is a margin of safety adequate, and what does it

mean to "protect" public health, given that any air pollution would have some effect on humans? A more subtle misuse of science arises under the Endangered Species Act, in its instruction to listing agencies to determine the suitability of a species for listing based only on science and purely factual data. 16 U.S.C. § 1533(b)(1)(A) (2012). The problem again is that the standards being applied require normative clarification. At what point is a species "in danger of extinction" in terms of percentage chance of disappearance over a given number of years? As for the particular factors pushing a species toward the edge, how is a scientist to decide whether regulatory protections are adequate and whether the species is being not just used but overused? As Professor Holly Doremus has explained, this science-based process cannot be carried out with any real predictability unless we get clear on the policy reasons for protecting species. Doremus, "Listing Decisions Under the Endangered Species Act: Why Better Science Isn't Always Better Policy," *Washington University Law Quarterly* 75 (1997): 1029.

25. See William Cronon, "The Trouble with Wilderness; or, Getting Back to the Wrong Nature," in *Uncommon Ground*, 69, 81–83 (presenting the controversial claim that the environmental movement was too influenced by an ideal of untouched nature).

26. The work of the historian Samuel P. Hays is perhaps the best guide to the policy stances of the environmental movement over the past sixty years: see *A History of Environmental Politics since 1945* (Cambridge: Cambridge University Press, 2000), 26–28, 51, 200; also see *Beauty, Health, and Permanence: Environmental Politics in the United States 1955–1985* (Cambridge: Cambridge University Press, 1987), 52–56, 123–30.

27. 33 U.S.C. § 1251(a) (2012). The other law that arguably employs a similarly strict standard is the food safety rule known as the Delaney Amendment, setting a zero tolerance for carcinogens in food for human consumption. Food Additives Transitional Provisions Amendment of 1961, 21 U.S.C. § 348(c)(3)(A) (2012). This is not a land-use rule, however, and the food itself can be highly unnatural in the sense that it reflects artificial breeding, production, and processing.

28. The Endangered Species Act protects rare species, investing them with intrinsic value, but it is not nearly the powerful tool that it is so often termed. That view derives from a misunderstanding of the narrow application of Tenn. Valley Authority v. Hill, 437 U.S. 153, 172, 195 (1978), a dispute involving a set of facts—a single government action that would eliminate a species—that has so far been unique. In practice, habitat loss can usually continue even when it elevates the danger of extinction. The statute's qualified protections are surveyed in Eric T. Freyfogle and Dale D. Goble, *Wildlife Law: A Primer* (Washington, DC: Island Press, 2009), 257–71.

29. See Cronon, "Trouble with Wilderness," 69, 81–83 (discussing the paradox by which human management of wilderness is necessary for wilderness protection); see also Joseph R. Desjardins, *Environmental Ethics: An Introduction to Environmental Philosophy*, 4th ed. (Belmont, CA: Wadsworth, 2006), 159–60 ("The suggestion is that if we simply leave it alone, the wilderness will be preserved in all its natural, unspoiled wonder. . . . But this assumption has problems of its own. Few areas on earth are unaffected by human activity").

30. Aldo Leopold, "Wilderness as a Form of Land Use," in *River of the Mother of God*, ed. Flader and Callicott, 136.

31. Sigurd F. Olson, *The Singing Wilderness* (New York: Alfred A. Knopf, 1956), 199–200, 208.

32. Robert A. Croker, *Pioneer Ecologist: The Life and Work of Victor Ernest Shelford, 1877–*

1968 (Washington, DC: Smithsonian Institution Press, 1991), 120–54 (surveying Shelford's nature preservation work).

33. See Eric T. Freyfogle, "What Is Good Land Use?," in Freyfogle, *Why Conservation Is Failing*, 144–77.

34. 16 U.S.C. § 1133 (2012); 16 U.S.C § 668dd (2014). See also Eric T. Freyfogle, "The Wildlife Refuge and the Land Community," *Natural Resources Journal* 44 (2004): 1027 (discussing the wildlife refuge context, and developing the need for refuge advocates to explain how the refuges benefit larger landscapes ecologically).

35. Sean Kammer, "Coming to Terms with Wilderness: The Wilderness Act and the Problem of Wildlife Restoration," *Environmental Law* 43 (2013): 83, 84–85.

36. See, e.g., S. T. A. Pickett et al., "Urban Ecological Systems: Scientific Foundations and a Decade of Progress," *Journal of Environmental Management* 92 (2011): 331; Marina Alberti, "The Effects of Urban Patterns on Ecosystem Function," *International Regional Science Review* 28 (2005): 168; R. W. Howarth, "Human Acceleration of the Nitrogen Cycle: Drivers, Consequences, and Steps Toward Solutions," *Water Science and Technology* 49 (2004): 7–12.

37. In consider the point in Eric. T. Freyfogle, "Life in the Enclaves," in Freyfogle, *Agrarianism and the Good Society: Land, Culture, Conflict, and Hope* (Lexington: University Press of Kentucky, 2007), 9, 17, 22.

38. Aldo Leopold, "Wilderness," in Leopold, *Sand County Almanac*, 188.

39. See, e.g., Curt Meine, "Building the 'Land Ethic,'" in *Companion to "A Sand County Almanac,"* ed. Callicott, 172, 183.

40. Dennis Ribbens, "The Making of *A Sand County Almanac*," in ibid., 91, 107.

41. See Aldo Leopold, "Sand County Almanac Manuscript" (1948), in Aldo Leopold Archives (University of Wisconsin Digital Collection), available http://digital.library.wisc.edu/1711.dl/AldoLeopold.ALSandCounty.

42. Leopold, *Sand County Almanac*, 200–201, 226.

43. Howard Zahniser, "The Need for Wilderness Areas," *The Living Wilderness* 59 (1956): 37, 38. I thank Peter Landres of the Aldo Leopold Wilderness Research Institute for drawing my attention to this writing. See 38–43 (advocating a sustainable societal connected with wilderness).

CHAPTER SEVEN

1. Garrett Hardin, "The Tragedy of the Commons," *Science* 162 (1968): 1243. A hint of the utility of the article can be gained by a search for citations to it in the Westlaw data base alone; a search on December 11, 2014, turned up 1,855 items.

2. Ibid., 1244.

3. A classic discussion is Siegfried V. Ciriacy-Wantrup and Richard C. Bishop, "'Common Property' as a Concept in Natural Resource Policy," *Natural Resources Journal* 15 (1975): 713.

4. Hardin, "Tragedy of the Commons," 1247. Hardin emphasized this point to counter any implication that "coercion implie[d] arbitrary decisions of distant and irresponsible bureaucrats."

5. Ibid., 1245 ("The tragedy of the commons as a food basket is averted by private property, or something formally like it").

6. Ibid., 1247. Taken as a whole, Hardin's article treats privatization as a form of mutual coercion, as it is. But in early parts of his article, before introducing the idea of mutual coercion, he presents private property as a solution without describing it in those terms, thus facilitating the view that they are different options (1245) ("The tragedy

of the commons as a food basket is averted by private property, or something formally like it"; to avert tragedy of commons in National Parks "we might sell them off as private property").

7. The distinction is explained and applied in Mark Sagoff, *The Economy of the Earth: Philosophy, Law, and the Environment* (Cambridge: Cambridge University Press, 1988), 7–8, 51–55, 65–67. Earlier considerations, many in the field of welfare economics, include Stephen Marglin, "The Social Rate of Discount and the Optimal Rate of Investment," *Quarterly Journal of Economics* 77 (1963): 95.

8. Hardin, "The Tragedy of the Commons," 243 ("A finite world can support only a finite population").

9. The term "commons" is explained in Ciriacy-Wantrup and Bishop, "'Common Property' and as a Concept."

10. On fisheries, see Jonathan H. Adler and Nathaniel Stewart, "Learning How to Fish: Catch Shares and the Future of Fisheries Conservation," *UCLA Journal of Environmental Law and Policy* 31 (2013): 150. In the case of the atmosphere, Mary Christina Wood has urged that it be embraced legally as a public trust asset with state duties to protect it in the public interest; Wood, *Nature's Trust*. Hardin viewed the atmosphere in the same way, realized that it could not be divided into private shares, and urged that mutual coercion by applied to protect it; Hardin, "The Tragedy of the Commons," 1245, 1247.

11. Leading studies of commons that have been well-managed, sometimes for centuries, include Ostrom, *Governing the Commons*, and Margaret A. McKean, "Success on the Commons: A Comparative Examination of Institutions for Common Property Resources Management," *Journal of Theoretical Politics* 4 (1992): 247. A positive view of the commons is presented in Anna di Robilant, "The Virtues of Common Ownership," *Boston University Law Review* 91 (2011): 1359.

12. Hardin, "The Tragedy of the Commons," 1247.

13. For instance, Ostrom, *Governing the Commons*, 12–13, *passim*.

14. Hardin, "The Tragedy of the Commons," 1243. In addition, Hardin made clear that landscapes divided into shares of privately owned land remained a commons with respect to water and air that carried away pollution (1245).

15. The urban setting is considered in Nicole Stelle Garnett, "Managing the Urban Commons," *University of Pennsylvania Law Review* 160 (2012): 1995, and Sheila R. Foster, "Collective Action and the Urban Commons," *Notre Dame Law Review* 87 (2011): 57. Garnett and Foster pay attention chiefly to the urban spaces that are not privately owned. My example considers a city as a whole without regard for ownership of particular parts; the city as such remains a commons given the interconnection of its parts and the ways the conduct of one actor affects others. A related line of scholarship considers municipalities as actors within larger spatial scales. For instance, Jonathan Rosenbloom, "New Day at the Pool: State Preemption, Common Pool Resources, and Non-Place Based Municipal Collaborations," *Harvard Environmental Law Review* 36 (2012): 445.

16. The typical tools to reduce the waste of too many wells are rules limiting well spacing and setting minimum tract sizes for each well, together with unitization procedures that allow for the management of entire fields as single production operations. Walter L. Summers, *The Law of Oil and Gas*, vol. 1 and 2014 Supp. (Eagan, MN: Thomson/West, 2004), 267–72.

17. Texas is among the states that have sought to reduce inefficiency in groundwater extraction—and to conserve water in the process—in part by using well-spacing,

well-production, and other rules long common in oil and gas fields. See Heather Welles, Note, "Toward a Management Doctrine for Texas Groundwater," *Ecology Law Quarterly* 40 (2013): 483, 491–94.

18. Privatization as the preferred solution is critically considered in Ostrom, *Governing the Commons*, 12–15, and Amy Sinden, "The Tragedy of the Commons and the Myth of a Private Property Solution," *University of Colorado Law Review* 78 (2007): 553.

19. A vivid and thoughtfully probed historical inquiry is Donald Worster, *The Dust Bowl: The Southern Plains in the 1930s* (New York: Oxford University Press, 1979), exploring the cultural origins of misuses of private lands.

20. On protecting wildlife populations the science and policy challenges of protecting species, particularly imperiled ones, are considered in Reed F. Noss and Allen Y. Cooperrider, *Saving Nature's Legacy: Protecting and Restoring Biodiversity* (Washington, DC: Island Press, 1994). One useful inquiry regarding controlling excessive drainage or land-cover change is David K. Mears and Sarah McKearnan, "Rivers and Resilience: Lessons Learned from Tropical Storm Irene," *Vermont Journal of Environmental Law* 14 (2012): 177. One thoughtful study on managing river floodplains is J. B. Ruhl et al., "Proposal for a Model State Watershed Management Act," *Environmental Law* 33 (2003): 929.

21. Hardin, "The Tragedy of the Commons," 1247.

22. Scholarly works strongly favoring privatization include Adler and Stewart, "Learning How to Fish," (for fisheries management), and Jan G. Laitos and Rachel B. Gamble, "The Problem with Wilderness," *Harvard Environmental Law Review* 32 (2008); 503 (for wilderness areas).

23. Joseph William Singer and Jack M. Beerman, "The Social Origins of Property," *Canadian Journal of Law and Jurisprudence* 6 (1993): 217.

24. I consider options available to lawmakers when crafting property systems in Freyfogle, *The Land We Share*, 1–36.

25. A concise summary of the issues appears in Eric T. Freyfogle and Bradley C. Karkkainen, *Property Law: Power, Governance, and the Common Good* (Eagan, MN: Thomson/West, 2012), 393–96; see also 312–14 for the options for resolving disputes among owners.

26. The ideas set forth in this paragraph and the next are developed further in Eric T. Freyfogle, "Good-bye to the Public-Private Divide," in *Agrarianism and the Good Society: Land, Culture, Conflict, and Hope* (Lexington: University Press of Kentucky, 2007), 83–106. The line is also blurred considerably in proposals for laws that facilitate private action to govern particular neighborhoods or landscapes. See, e.g., Robert C. Ellickson, "New Institutions for Old Neighborhoods," *Duke Law Journal* 48 (1998): 75; Robert H. Nelson, "Privatizing the Neighborhood: A Proposal to Replace Zoning with Private Collective Property Rights to Existing Neighborhoods," *George Mason Law Review* 7 (1999): 827. Adding further complexity is the frequent existence of multiple levels of governance. Blake Hudson, "Federal Constitutions: The Keystone of Nested Commons Governance," *Alabama Law Review* 63 (2012): 1007.

27. I explore the challenge in *Our Oldest Task: Making Sense of Our Place in Nature* (Chicago: University of Chicago Press, 2017).

28. Hardin's examples all qualify as actions that are harmful only when too many people engage in them.

29. A useful consideration of options is Christopher S. Elmendorf, "Ideas, Incentives, Gifts, and Governance: Toward Conservation Stewardship of Private Land in Cultural and Psychological Perspective," *University of Illinois Law Review* (2003): 423.

30. Freyfogle, *The Land We Share*, 203–27.
31. I challenge this perspective in Freyfogle, *On Private Property*, 12–26.
32. The challenges posed by nature itself are considered in Bradley C. Karkkainen, "Collaborative Ecosystem Governance: Scale, Complexity, and Dynamism," *Virginia Environmental Law Journal* 21 (2002): 189.
33. A classic historical study of the dialectical interaction of nature and culture, and how it plays out in terms of property-use systems, is William Cronon, *Changes in the Land: Indians, Colonists, and the Ecology of New England* (New York: Hill and Wang, 1983).
34. These challenges and many others are considered in Boyce Thorne-Miller, "Setting the Rights Goals: Marine Fisheries and Sustainability in Large Ecosystems," in *Precautionary Tools for Reshaping Environmental Policy*, ed. Nancy J. Myers and Carolyn Raffensperger (Cambridge, MA: MIT Press, 2006), 155.
35. I consider the similarities and identify differences in *The Land We Share*, 174–77.
36. This view of human behavior is explored and criticized in Russell B. Korobkin and Thomas S. Ulen, "Law and Behavioral Science: Removing the Rationality Assumption from Law and Economics," *California Law Review* 88 (2000): 1051.
37. The question as posed implicitly raises foundational questions: Why do humans degrade nature? What are the root causes of our misguided acts? The literature on the subject is far less developed than it should be given the importance of the questions and the need ultimately for solutions that address root causes. Thoughtful comments on our overall plight are offered in Worster, *The Wealth of Nature*. A critique of market capitalism as a cause of degradation is presented in Fred Magdoff and John Bellamy Foster, *What Every Environmentalist Needs to Know about Capitalism* (New York: Monthly Review Press, 2011).
38. Hardin introduced this line of reasoning in the context of population growth to explain why the challenge could not be met simply by appealing to the consciences of individuals as such. Hardin, "The Tragedy of the Commons," 1246–47.
39. The various strands of ethical thought relating to our uses of nature are considered and synthesized in a dated but still useful source, Bryan G. Norton, *Toward Unity among Environmentalists* (New York: Oxford University Press, 1991).
40. Freyfogle, *Justice and the Earth*, 26–42.
41. The argument is developed in Magdoff and Foster, *What Every Environmentalist*, and Chris Williams, *Ecology and Socialism: Solutions to Capitalist Ecological Crisis* (Chicago: Haymarket Books, 2010), 226 ("It is the economic system that dictates that nonsustainability is rational, not people").
42. A thoughtful inquiry is Barton H. Thompson Jr., "Tragically Difficult: The Obstacles to Governing the Commons," *Environmental Law* 30 (2000): 241.
43. Such fears might be well grounded in the realities of the settings in which interested parties come together. Discussions could be skewed to favor particular interests that wield disproportionate power. Amy Sinden, "In Defense of Absolutes: Combatting the Politics of Power in Environmental Law," *Iowa Law Review* 90 (2005): 1405. Similarly, the processes of engagement could be ones that are so drawn out, and so costly for the participants, that only the best-funded interests can stay engaged. John D. Echeverria, "No Success Like Failure: The Platte River Collaborative Watershed Planning Process," *William and Mary Environmental Law and Policy Review* 25 (2001): 559. More generally, cooperation that takes place through institutions is subject to the limitations of such institutions; the topic is surveyed in Daniel H. Cole, "The Varieties of Comparative Institutional Analysis," *Wisconsin Law Review* (2013): 383.
44. Williams, *Ecology and Socialism* 43–44.

45. This point, and those made in the next three sentences, appear prominently in the many writings of Wendell Berry, e.g., "Conservation and Local Economy" in Berry, *Sex, Economy, Freedom and Community*, 3–18.

46. Beyond these points, users of the commons would typically need to develop governance arrangements that responded to the various forces that could undercut successful commons management. The traits of successful commons-management arrangements are surveyed in Ostrom, *Governing the Commons*, 58–102. Ostrom usefully probes the complex motives of individual users of commons in "A Behavioral Approach to the Rational Choice Theory of Collective Action," *American Political Science Review* 92 (1998): 1.

47. Wendell Berry, "Private Property and the Common Wealth," in Berry, *Another Turn of the Crank*, 46–63.

48. The issue is considered in John D. Echeverria, "Regulating Versus Paying Land Owners to Protect the Environment," *Journal of Land Resources and Environmental Law* 26 (2005): 1.

49. A vibrant civil society is likely also a prerequisite. Julianne Lutz Newton and William C. Sullivan, "Nature, Culture, and Civil Society," *Journal of Civil Society* 1 (2005): 195.

50. The volume remains in print from various publishers. It originally appeared in differing editions between 1714 and 1732.

51. The argument is made by various writers including Magdoff and Foster, *What Every Environmentalist Needs to Know about Capitalism*, and Williams, *Ecology and Socialism*.

52. Donald Worster, *Shrinking the Earth: The Rise and Decline of Nature's Abundance* (New York: Oxford University Press, 2016), 5–8.

53. Alfred E. Kahn, "The Tyranny of Small Decisions: Market Failures, Imperfections, and the Limits of Economics," *Kyklos* 19 (1966): 23.

54. William E. Odum, "Environmental Degradation and the Tyranny of Small Decisions," *BioScience* 32 (1982): 728.

55. Hardin, "The Tragedy of the Commons," 1245.

56. I use the term and elaborate the idea in Freyfogle, *The Land We Share*, 157–78.

57. As noted in chapter 1, the point served as a central theme of Leopold's *Almanac*, beginning with two much-quoted sentences from the book's foreword: "We abuse land because we regard it as a commodity belonging to us. When we see land as a community to which we belong, we may begin to use it with love and respect." Leopold, *Sand County Almanac*, viii.

58. Wendell Berry makes the point, which is central to his writings, in "Health Is Membership," 86.

CONCLUSION

1. Mark Shepard, *Restoration Agriculture: Real-World Permaculture for Farmers* (Austin, TX: Acres U.S.A., 2013).

2. Louis Hartz, *The Liberal Tradition in America* (New York: Harcourt, Brace, 1955).

3. Ostrom, *Governing the Commons*.

4. Jeremy Bentham, *The Theory of Legislation* (Chap. VIII, "On Property") (1811), quoted in C. B. Macpherson, *Property: Mainstream and Critical Positions* (Toronto: University of Toronto Press, 1978), 52.

5. Cronon, *Changes in the Land*.

6. Bellah et al., *Habits of the Heart*.

SELECTED BIBLIOGRAPHY

Alexander, Gregory S. *Commodity and Propriety: Competing Visions of Property in American Legal Thought, 1776–1970*. Chicago: University of Chicago Press, 1997.

Alperovitz, Gar. *America beyond Capitalism: Reclaiming Our Wealth, Our Liberty, and Our Democracy*. Hoboken, NJ: John Wiley and Sons, 2005.

Barber, Benjamin R. *Strong Democracy: Participatory Politics for a New Age*. 20th anniv. ed. Berkeley: University of California Press, 2003.

Beatley, Timothy. *Ethical Land Use: Principles of Policy and Planning*. Baltimore: Johns Hopkins University Press, 1994.

Bell, Daniel. *The Cultural Contradictions of Capitalism*. 20th anniv. ed. New York: Basic Books, 1996.

Bellah, Robert N., et al. *Habits of the Heart: Individualism and Commitment in American Life*. Updated ed. Berkeley: University of California Press, 1996.

Berry, Thomas. *The Dream of the Earth*. San Francisco: Sierra Club Books, 1988.

Berry, Wendell. *Another Turn of the Crank*. Berkeley, CA: Counterpoint, 1995.

———. *A Continuous Harmony: Essays Cultural and Agricultural*. San Diego: Harcourt, Brace, 1972.

———. *The Gift of Good Land*. San Francisco: North Point Press, 1981.

———. *Home Economics*. San Francisco: North Point Press, 1987.

———. *Sex, Economy, Freedom, and Community*. New York: Pantheon, 1992.

———. "The Total Economy." In Berry, *Citizenship Papers*. Washington, DC: Shoemaker and Hoard, 2003.

———. *The Unsettling of America: Culture and Agriculture*. San Francisco: Sierra Club Books, 1977.

———. *The Way of Ignorance and Other Essays*. Emeryville, CA: Shoemaker & Hoard, 2005.

———. *What Are People For?* San Francisco: North Point Press, 1990.

Bollier, David. *Silent Theft: The Private Plunder of Our Common Wealth*. New York: Routledge, 2003.

Brinkley, Alan. *Liberalism and Its Discontents*. Cambridge, MA: Harvard University Press, 1998.

Callicott, J. Baird, ed. *Companion to "A Sand County Almanac": Interpretive and Critical Essays*. Madison: University of Wisconsin Press, 1987.

———. "Ecological Sustainability as a Conservation Concept." In Callicott, *Beyond the Land Ethic: More Essays in Environmental Philosophy*. Albany, NY: SUNY Press, 1999.

———. *In Defense of the Land Ethic: Essays in Environmental Philosophy.* Albany: SUNY Press, 1989.

Chang, Ha-Joon. *23 Things They Don't Tell You about Capitalism.* New York: Bloomsbury, 2011.

Daly, Herman E., and John B. Cobb Jr. *For the Common Good: Redirecting the Economy Toward Community, the Environment, and a Sustainable Future.* Boston: Beacon Press, 1989.

Eckersley, Robyn. *Environmentalism and Political Theory: Toward an Ecocentric Approach.* Albany: SUNY Press, 1992.

———. *The Green State: Rethinking Democracy and Sovereignty.* Cambridge, MA: MIT Press, 2004.

Ehrenfeld, David. *The Arrogance of Humanism.* New York: Oxford University Press, 1978.

———. *Swimming Lessons: Keeping Afloat in the Age of Technology.* New York: Oxford University Press, 2002.

Etzioni, Amitai. *The New Golden Rule: Community and Morality in a Democratic Society.* New York: Basic Books, 1996.

Evernden, Neil. *The Natural Alien: Humankind and Environment.* Toronto: University of Toronto Press, 1985.

Fiege, Mark. *The Republic of Nature: An Environmental History of the United States.* Seattle: University of Washington Press, 2012.

Fawcett, Edmund. *Liberalism: The Life of an Idea.* Princeton, NJ: Princeton University Press, 2014.

Flader, Susan L. *Thinking Like a Mountain: Aldo Leopold and the Evolution of an Ecological Attitude Toward Deer, Wolves, and Forests.* Columbia: University of Missouri Press, 1974.

Flader, Susan L., and J. Baird Callicott, eds. *The River of the Mother of God and Other Essays by Aldo Leopold.* Madison: University of Wisconsin Press, 1991.

Foster, John Bellamy. *The Ecological Revolution: Making Peace with the Planet.* New York: Monthly Review Press, 2009.

Fowler, Robert Booth. *The Greening of Protestant Thought.* Chapel Hill: University of North Carolina Press, 1995.

Frank, Robert H. *The Darwin Economy: Liberty, Competition, and the Common Good.* Princeton, NJ: Princeton University Press, 2011.

Freyfogle, Eric T., ed. *Agrarianism and the Good Society: Land, Culture, Conflict, and Hope.* Lexington: University Press of Kentucky, 2007.

———. *Justice and the Earth: Images for Our Planetary Survival.* New York: Free Press, 1993.

———. *The Land We Share: Private Property and the Common Good.* Washington, DC: Island Press, 2003.

———. *On Private Property: Finding Common Ground on the Ownership of Land.* Boston: Beacon Press, 2007.

———. *Our Oldest Task: Making Sense of Our Place in Nature.* Chicago: University of Chicago Press, 2017.

———. *Why Conservation Is Failing and How It Can Regain Ground.* New Haven, CT: Yale University Press, 2006.

Freyfogle, Eric T., and Dale D. Goble. *Wildlife Law: A Primer.* Washington, DC: Island Press, 2009.

Freyfogle, Eric T., and Bradley C. Karkkainen. *Property Law: Power, Governance, and the Common Good.* Eagan, MN: Thomson/West, 2012.

Genovese, Eugene D. *The Southern Tradition: The Achievement and Limitations of an American Conservatism.* Cambridge, MA: Harvard University Press, 1994.

Goodrich, Janet. *The Unforeseen Self in the Works of Wendell Berry.* Columbia: University of Missouri Press, 2001.

Gray, John. *Enlightenment's Wake: Politics and Culture at the Close of the Modern Age.* London: Routledge, 1995.

Hacker, Jacob S., and Paul Pierson. *Winner-Take-All Politics: How Washington Made the Rich Richer—and Turned Its Back on the Middle Class.* New York: Simon and Schuster, 2010.

Harari, Yuval Noah. *Sapiens: A Brief History of Humankind.* London: Harvill Secker, 2014.

Hays, Samuel P. *Beauty, Health, and Permanence: Environmental Politics in the United States, 1955–1985.* Cambridge: Cambridge University Press, 1987.

———. *A History of Environmental Politics since 1945.* Pittsburgh: University of Pittsburgh Press, 2000.

Jackson, Wes. *New Roots for Agriculture.* Rev. ed. Lincoln: University of Nebraska Press, 1985.

James, William. *Pragmatism: A New Name for Some Old Ways of Thinking.* New York: Longmans, Green and Co., 1907.

Johnson, Mark. *Morality for Humans: Ethical Understanding from the Perspective of Cognitive Science.* Chicago: University of Chicago Press, 2014.

Joyce, Richard. *The Evolution of Morality.* Cambridge, MA: MIT Press, 2006.

Krutch, Joseph Wood. *The Modern Temper: A Study and a Confession.* New York: Harcourt, Brace and Co., 1929.

Lears, Jackson. *Rebirth of a Nation: The Making of Modern America, 1877–1920.* New York: Harper, 2009.

Leopold, Aldo. *For the Health of the Land: Previously Unpublished Essays and Other Writings.* Edited by J. Baird Callicott and Eric T. Freyfogle. Washington, DC: Island Press, 1999.

———. *A Sand County Almanac and Sketches Here and There.* New York: Oxford University Press, 1949.

Leopold, Luna, ed. *Round River: From the Journals of Aldo Leopold.* New York: Oxford University Press, 1953.

Lieberman, Matthew D. *Social: Why Our Brains Are Wired to Connect.* New York: Broadway Books, 2013.

MacIntyre, Alasdair. *After Virtue.* 2nd ed. Notre Dame, IN: University of Notre Dame Press, 1984.

Magdoff, Fred, and John Bellamy Foster. *What Every Environmentalist Needs to Know about Capitalism: A Citizen's Guide to Capitalism and the Environment.* New York: Monthly Review Press, 2011.

McNeill, J. R. *Something New Under the Sun: An Environmental History of the Twentieth-Century World.* New York: W. W. Norton, 2000.

Meine, Curt. *Aldo Leopold: His Life and Work.* Madison: University of Wisconsin Press, 1988.

———. *Corrections Lines: Essays on Land, Leopold, and Conservation.* Washington, DC: Island Press, 2004.

Menand, Louis. *The Metaphysical Club: A Story of Ideas in America.* New York: Farrar, Straus, Giroux, 2001.

Mill, John Stuart. *On Liberty.* 1859; Indianapolis, IN: Bobbs-Merrill, 1956.

———. *Utilitarianism.* 1861; New York: Barnes and Noble, 2005.

Miller, Char. *Gifford Pinchot and the Making of Modern Environmentalism.* Washington, DC: Island Press, 2001.

Mumford, Lewis. *Technics and Human Development.* New York: Harcourt Brace Jovanovich, 1967.

Nash, Roderick. *Wilderness and the American Mind.* 4th ed. New Haven, CT: Yale University Press, 2001.

Newton, Julianne Lutz. *Aldo Leopold's Odyssey: Rediscovering the Author of "A Sand County Almanac."* Washington, DC: Island Press, 2006.

Northcott, Michael S. *A Moral Climate: The Ethics of Global Warming.* London: Darton, Longman and Todd, 2007.

Norton, Bryan G. *Searching for Sustainability: Interdisciplinary Essays in the Philosophy of Conservation Biology.* Cambridge: Cambridge University Press, 2003.

Noss, Reed F., and Allen Y. Cooperrider. *Saving Nature's Legacy: Protecting and Restoring Biodiversity.* Washington, DC: Island Press, 1994.

Oelschlaeger, Fritz. *The Achievement of Wendell Berry: The Hard History of Love.* Lexington: University Press of Kentucky, 2011.

Oelschlaeger, Max. *The Idea of Wilderness: From Prehistory to the Age of Ecology.* New Haven, CT: Yale University Press, 1991.

Ophuls, William. *Plato's Revenge: Politics in the Age of Ecology.* Cambridge, MA: MIT Press, 2011.

———. *Requiem for Modern Politics: The Tragedy of the Enlightenment and the Challenge of the New Millennium.* Boulder, CO: Westview Press, 1997.

Orr, David W. *Down to the Wire: Confronting Climate Collapse.* New York: Oxford University Press, 2009.

———. *Earth in Mind: On Education, Environment, and the Human Prospect.* Washington, DC: Island Press, 1994.

———. *Ecological Literacy: Education and the Transition to a Postmodern World.* Albany: SUNY Press, 1992.

———. *Hope Is an Imperative: The Essential David Orr.* Washington, DC: Island Press, 2011.

———. *The Last Refuge: Patriotism, Politics, and the Environment in an Age of Terror.* Washington, DC: Island Press, 2004.

———. *Nature by Design: Ecology, Culture, and Human Intention.* New York: Oxford University Press, 2002.

Ostrom, Elinor. *Governing the Commons: The Evolution of Institutions for Collective Action.* Cambridge: Cambridge University Press, 1990.

Peters, Jason, ed. *Wendell Berry: Life and Work.* Lexington: University Press of Kentucky, 2007.

Polanyi, Karl. *The Great Transformation: The Political and Economic Origins of Our Time.* 1944; Boston: Beacon Press, 1957.

Ponting, Clive. *A New Green History of the World: The Environment and the Collapse of Great Civilizations.* Rev. ed. New York: Penguin Books, 2007.

Rodgers, Daniel T. *Age of Fracture.* Cambridge, MA: Harvard University Press, 2011.

———. *Contested Truths: Keywords in American Politics since Independence.* New York: Basic Books, 1987.

Sagoff, Mark. *The Economy of the Earth: Philosophy, Law, and the Environment.* Cambridge: Cambridge University Press, 1988.

Schlatter, Richard. *Private Property: The History of an Idea.* London: George Allen and Unwin, 1951.

Scott, William B. *In Pursuit of Happiness: American Conceptions of Property from the Seventeenth to the Twentieth Century.* Bloomington: Indiana University Press, 1977.

Scoville, J. Michael. "Environmental Values, Animals, and the Ethical Life." PhD dissertation, Order No. 3496684, University of Illinois at Urbana-Champaign. Ann Arbor, MI: ProQuest, 2011.

Siedentop, Larry. *Inventing the Individual: The Origins of Western Liberalism*. Cambridge, MA: Belknap Press, 2014.

Sen, Amartya. *On Ethics and Economics*. Oxford: Basil Blackwell, 1987.

Shapiro, Ian. *The Moral Foundations of Politics*. New Haven, CT: Yale University Press, 2003.

Singer, Peter. *The Expanding Circle: Ethics, Evolution, and Moral Progress*. Rev. ed. Princeton, NJ: Princeton University Press, 2011.

Smith, Kimberly K. *Wendell Berry and the Agrarian Tradition: A Common Grace*. Lawrence: University of Kansas Press, 2003.

Steinberg, Ted. *Down to Earth: Nature's Role in American History*. 2nd ed. New York: Oxford University Press, 2009.

Tarnas, Richard. *The Passion of the Western Mind: Understanding the Ideas that Have Shaped Our World View*. New York: Crown Publishers, 1991.

Tawney, R. H. *The Acquisitive Society*. London: G. Bell and Sons, 1921.

Thomas, Keith. *Man and the Natural World: A History of the Modern Sensibility*. New York: Pantheon, 1983.

Tomasello, Michael. *A Natural History of Human Thinking*. Cambridge, MA: Harvard University Press, 2014.

Watson, Peter. *The Modern Mind: An Intellectual History of the Twentieth Century*. New York: Harper Collins, 2001.

Watts, Alan W. *Nature, Man and Woman*. New York: Pantheon, 1958.

Weston, Burns H., and David Bollier. *Green Governance: Ecological Survival, Human Rights, and the Law of the Commons*. New York: Cambridge University Press, 2013.

Williams, Chris. *Ecology and Socialism: Solutions to Capitalist Ecological Crisis*. Chicago: Haymarket Books, 2010.

Wilson, Edward O. *The Social Conquest of Earth*. New York: Liveright Publishing, 2012.

Wirzba, Norman, ed. *The Art of the Commonplace: The Agrarian Essays of Wendell Berry*. Washington, DC: Counterpoint Press, 2002.

Wood, Mary Christina. *Nature's Trust: Environmental Law for a New Ecological Age*. New York: Cambridge University Press, 2014.

Worster, Donald. *Dust Bowl: The Southern Plains in the 1930s*. New York: Oxford University Press, 1979.

———. *Nature's Economy: A History of Ecological Ideas*. 2nd ed. Cambridge: Cambridge University Press, 1994.

———. *A Passion for Nature: The Life of John Muir*. New York: Oxford University Press, 2008.

———. *Shrinking the Earth: The Rise and Decline of American Abundance*. New York: Oxford University Press, 2016.

———. *The Wealth of Nature: Environmental History and the Ecological Imagination*. New York: Oxford University Press, 1993.